Broadband Quantum Cryptography

Synthesis Lectures on Quantum Computing

Editors
Marco Lanzagorta, ITT Corporation
Jeffrey Uhlmann, University of Missouri-Columbia

Broadband Quantum Cryptography

Daniel J. Rogers

ISBN: 978-3-031-01385-0   paperback
ISBN: 978-3-031-02513-6   ebook

DOI 10.1007/978-3-031-02513-6

A Publication in the Springer series
SYNTHESIS LECTURES ON QUANTUM COMPUTING

Lecture #3
Series Editors: Marco Lanzagorta, ITT Corporation
                Jeffrey Uhlmann, University of Missouri-Columbia
Series ISSN
Synthesis Lectures on Quantum Computing
Print 1945-9726   Electronic 1945-9734

Broadband Quantum Cryptography

Daniel J. Rogers
Johns Hopkins University Applied Physics Laboratory

SYNTHESIS LECTURES ON QUANTUM COMPUTING #3

ABSTRACT

Quantum cryptography is a rapidly developing field that draws from a number of disciplines, from quantum optics to information theory to electrical engineering. By combining some fundamental quantum mechanical principles of single photons with various aspects of information theory, quantum cryptography represents a fundamental shift in the basis for security from numerical complexity to the fundamental physical nature of the communications channel. As such, it promises the holy grail of data security: theoretically unbreakable encryption. Of course, implementing quantum cryptography in real broadband communications systems poses some unique challenges, including generating single photons, distilling random keys from the quantum key distribution process, and maintaining security at both the theoretical and practical level. Overall, quantum cryptography has a place in the history of secret keeping as a novel and potentially useful paradigm shift in the approach to broadband data encryption.

KEYWORDS

quantum cryptography, quantum key distribution

Contents

Acknowledgments

There are a number of people who contributed greatly to both the work of creating this book and to the gathering of the knowledge contained in it. First and foremost, I must thank Dr. Charles Clark and Dr. Michael Coplan for their invaluable guidance and support. I must also thank my editor, Dr. Marco Lanzagorta, and my publisher, Mr. Mike Morgan, for the opportunity to write this book. I owe a debt of gratitude to Joseph Haber, Kevin Parker, and Charles Goldblum for their support as well, and to Devon Goforth and Peter Chin for offering their valued opinions about the material presented here. In addition, I would never have been able to complete these pages were it not for all that I learned from Alessandro Restelli, Alex Ling, and Josh Bienfang.

Finally, I am indebted to my family for motivating me to continually strive for success and to Miriam for her constant support and inspiration.

Daniel J. Rogers
April 2010

Preface

Quantum cryptography is one of the broadest and most dynamic subjects in modern applied physics. It draws upon a myriad of advancements in modern science and engineering, from quantum physics to information theory to high-speed digital electronics, in order to solve one of the oldest problems in communications: Keeping secrets. This text attempts to introduce the reader to the fundamentals of the field with an emphasis on practical implementation in the modern broadband communications environment. It assumes a basic knowledge of classical optics and introductory quantum mechanics but is intended for generally knowledgeable readership. Chapter 1 introduces quantum cryptography, its motivation, and its place in the history of secure communications. Chapter 2 offers a brief review of classical cryptography in order to understand the fundamental differences provided by the quantum approach. Chapter 3 provides a brief review of classical and quantum optics. Chapter 4 introduces the fundamental principles of quantum cryptography based on the material reviewed in Chapter 3. Chapter 5 provides a very brief summary of the relevant topics in information theory. Chapter 6 discusses the various components of a quantum cryptography system, while Chapter 7 puts everything together with a number of overviews of typical implementations. Finally, Chapter 8 discusses how quantum cryptography might fit into the modern telecommunications infrastructure.

My intent is for the text to be clear and easily readable, with as little jargon and as few unexplained acronyms as practicable. It is by no means a complete treatment of any one subject, be it quantum optics, information theory, or any of the engineering disciplines. Perhaps the most useful part of the entire lecture is the bibliography; here I've attempted to collect as current a list of the seminal works relating to quantum cryptography as possible in such a rapidly changing field. Overall, however, I hope to provide an overview deep enough for a new entrant to be able to understand the literature to which I refer, as well as enough background to convey how interesting and exciting this field of quantum information really is. Good luck and enjoy!

Daniel J. Rogers
April 2010

CHAPTER 1

Introduction

"Three may keep a secret, if two of them are dead." -Ben Franklin

1.1 A BRIEF HISTORY OF KEEPING SECRETS

Secrets. They are the enticing subject of everything from political scandals to mystery novels. As much as honesty plays an integral role in society's mores, there is no denying that secrets have an equally, if not more, profound impact on history and human behavior. Whether we talk of Andre Malraux's pessimistic existentialism ("What is man? A miserable little pile of secrets.") or Nixon's Watergate scandal ("I am not a crook!"), secrets, their keeping, and their revelation lie beneath many of history's most important events.

The way that civilizations have guarded their secrets has had an equally dramatic history as the secrets themselves. Robert Churchhouse describes how the ancient Greeks would carve messages into the shaved scalps of their slaves and let the hair grow back to conceal the message contents [Churchhouse, 2002]. Though this primitive form of steganography is not the most humane or efficient means of obfuscation, examples like this represent civilization's first organized foray into information hiding. Roman general Julius Caesar employed what we might recognize as the first documented use of cryptography when he transmitted his military correspondence using a "shift-3" cipher, replacing each letter of the alphabet with the one three places ahead of it. While simple by today's standards, the cipher remained effective in the face of enemies who were often not even literate, let alone versed in the mathematics required to break such a code. It wasn't until Arab mathematicians developed the necessary frequency analysis techniques in the 9th century AD that Caesar's cipher was finally vulnerable to attack [Singh, 2000]. Since then, the code makers have been exploiting deeper and deeper mathematics to hide their data, while code breakers have applied more advanced techniques to each challenge in order to break those codes. Ever since then, cryptology and cryptography have been locked in this centuries-old cat and mouse game, all the while raising the stakes with more and more sensitive information at risk every day.

1.2 WHY QUANTUM CRYPTOGRAPHY?

Occam's Razer extolls the virtue of simplicity, and the field of cryptography is certainly not exempt. As encryption methods have gotten increasingly complex in recent years, one method remains an outlier in the world of cryptography: the Vernam Cipher [Vernam, 1926]. Invented in the 1920's by Bell Telephone engineer Gilbert Vernam, this profoundly simple cipher remains to this day the only one currently in existence that has a sound theoretical proof of unbreakable security [Shannon and Weaver, 1949].

The cipher is simple. Suppose a sender, often named Alice, has a message consisting of a sequence of digits (recall that Claude Shannon showed that all messages can be represented by a sequence of digits. In fact, all messages can be written as a sequence of binary digits, provided one uses enough of them). Alice then writes down a "key" consisting of a completely random sequence of digits exactly as long as her message. To encrypt the message, Alice simply adds, one by one, each message digit with its corresponding key digit (modulo the length of the alphabet, of course). Shannon proved that under these conditions absolutely no information is contained in the resulting ciphertext [Shannon and Weaver, 1949]. This means that, without knowing Alice's key, there is no way one would be able to extract any part of the original message from the encrypted transmission. Of course, this claim is subject to a number of rules. First of all, Alice's key sequence must be truly random because any correlations in the key would enable an attacker to begin to guess some parts of the key and subsequently start to break the coded message. Second, the key must be exactly as long as the message and must never be reused. If any part of the key were reused, then an attacker could examine repeating portions of the ciphertext and subtract the surmised key bits, revealing part or all of the message. Imagine that Alice is writing her key down on a pad of paper. After each transmission, she would need to remove the top sheet, discard it, and start over with a new key for each new message. This is where the Vernam cipher acquired its alternative name, the one time pad.

The problem with the one time pad comes when the transmitted ciphertext needs to be decrypted. The process is simple in theory; the recipient, often named Bob, takes Alice's key and subtracts each key digit with its corresponding ciphertext digit, revealing Alice's original message. Unfortunately for our heroic duo, Bob must acquire the exact same key sequence that Alice used to encrypt the message, a task that, in the end, proves as difficult as securely sending the message in the first place. For this reason, the One Time Pad has been considered of no practical use to the information security community, and various approximate implementations of it that attempt to circumvent this issue (RC4, for example) have proven insecure [Fluhrer *et al.*, 2001].

For the time being, public key cryptosystems such as RSA have solved the key distribution problem through the clever application of number theory [Rivest *et al.*, 1978]. We will see in the following chapter that all such systems rely on what are called one-way, or trapdoor, functions. These are mathematical operations that are simple to compute in one direction and difficult to perform in reverse. One example is the multiplication of prime numbers. It is relatively straightforward to program a classical computer (i.e., one based on the universal Turing machine model) to multiply two large prime numbers into what is called a semi-prime (a number with only two non-unity prime factors). By straightforward, one refers to the existence of an algorithm that is $\mathcal{O}(n^k)$ for some integer k. That is, it can be computed in linear or, at worst, higher-order polynomial time. However, current algorithms to factor such a semiprime number into its constituent primes operate in at best semi-exponential time. That is, they are of order $\mathcal{O}((1 - \epsilon)^n)$ for any positive $\epsilon < 1$ [Pomerance *et al.*, 1988]. Thus an attacker would need to acquire exponentially more computational power every time the key length (and hence the computational power applied to the encryption) was increased by a linear amount. This relationship between making and breaking public key ciphers allows the secret keepers to continually stay one step ahead of potential adversaries.

However, this advantage may not persist indefinitely. Prime factorization, along with other one-way operations such as discrete logarithms, falls into the complexity class of NP-complete problems, and it remains unproven that faster algorithms to solve these problems do not exist. The absence of this proof is both one of the most important mathematical challenges of modern times as well as the single greatest threat to modern information security. The potential existence of faster factoring algorithms immediately compromises the security of RSA and a host of other complexity-based cryptosystems. In fact, if one relaxes the requirement that the adversary's computation power be limited to classical, Turing-based computers, then RSA and other public-key systems can already be considered insecure. In 1994, Peter Shor showed that computers based on quantum mechanical bits (aptly named *quantum computers*) are able to factor large primes in polynomial running time [Shor, 1994]. Currently, the challenge in quantum computing is not how to use it to break encryption methods, but rather how to build one with enough bits to be useful. Once that problem is overcome, all of the world's seemingly secure data will be subject to compromise.

That vulnerability is where quantum cryptography or, more specifically, quantum key distribution becomes significant. As we will discover in the coming chapters, using the principles of quantum mechanics intrinsic to certain transmission media can allow Alice and Bob to exchange a truly random key of arbitrary length suitable for use with a one time pad. The security of their exchange is based on the physics of the transmission medium rather than any assumptions about computational complexity or an attacker's ability. Thus quantum key distribution, or QKD as it is commonly abbreviated, represents a fundamental paradigm shift in encryption methodologies away from computational methods to ones grounded in the physical transmission layer.

The following chapters will illuminate just how the sometimes counterintuitive principles of quantum mechanics achieve this seemingly miraculous achievement of unbreakability. They will demonstrate the first real-world, practical utility of fundamental quantum mechanics to solving a problem faced every day by modern citizens. They will also discuss the limits encountered when this ideal solution meets the practical problems of engineering. Like any technology, QKD is often limited by the technological capabilities of the components used in its implementation. Often, this leads to skepticism among the less informed as to the true value of the idea. This book will attempt to quell that skepticism while providing an honest assessment of some of the current candidates for fieldable implementation.

Quantum cryptography is a nascent field from the point of view of real-world communications. However, the payoff in terms of information security is unprecedented. With QKD, we may finally end the cat and mouse game that has been going on between the guardians of secrets and their enemies since the 9th century.

CHAPTER 2

Elements of Classical Cryptography

While we will primarily focus on the science and technology of quantum encryption, it behooves us to review some relevant concepts from classical cryptography for a number of reasons. Not only is the cryptography section of your local library likely many shelves away, but the concepts that comprise classical cryptography are equally fundamental in the quantum approach. Thus we will provide a cursory treatment of some of the most important concepts of number and information theory, public key and symmetric cryptosystems, and pseudorandom number generation.

2.1   PRIME NUMBERS AND PUBLIC-KEY ENCRYPTION

When developing any new technology, it is always important to scope out the competition. In the case of QKD, its most ubiquitous competitor in secure communications is the RSA cryptosystem, named after its originators, Rivest, Shamir, and Adleman [Rivest et al., 1978]. Currently, the industry standard, RSA encryption relies on some fundamental facts about prime numbers - namely that it is easier computationally to find and multiply two large prime numbers than to factor their product back into its constituent primes. The following section outlines how this process provides the ability to communicate securely[1].

2.1.1   THE RSA CRYPTOSYSTEM

Suppose our sender, Alice, has a sensitive message that she wants to send to a recipient, Bob, over some open channel. Bob knows that Alice wants to send her a message, and Alice knows that Bob really is whom he claims to be (the so-called *authentication* problem, to be discussed later). In order to facilitate the communication, Bob chooses two large prime numbers, p and q, which he can do with relative ease[2]. He then computes $N = p \cdot q$ and also chooses an integer e such that

$$\gcd (e, \varphi(N)) = 1, \tag{2.1}$$

---

[1]Throughout this section, I offer a number of rather deep fundamental concepts from number theory without proofs. For further insight, the appropriately skeptical reader is referred to [Kranakis, 1986].

[2]As far back as 300 BC, Euclid proved that there are an infinite number of primes, and in the late 19th century, Hadamad and De La Valleé showed that for any number n, there are approximately $\frac{n}{\ln n}$ prime numbers less than n. That is, prime numbers are actually more common than square numbers [Euclid, 1908].

where $\varphi(N)$ is the Euler totient function, defined as the number of integers greater than zero but less than N that are coprime[3] to N. Two interesting properties of the Euler φ function that will become important later: First, if p is a prime number, then $\varphi(p) = p - 1$. Second, if two numbers, m and n are coprime, that is gcd $(m, n) = 1$, then $\varphi(m \cdot n) = \varphi(m) \cdot \varphi(m)$ (i.e., φ is *weakly multiplicative*). Thus Bob essentially computes gcd$(e, (p-1) \cdot (q-1))$. He also computes $d = e^{-1}$, or more precisely the integer d such that

$$d \cdot e = 1 \quad \mathrm{mod} \ \varphi(N), \tag{2.2}$$

a choice that is computationally straightforward using Euclid's algorithm [Euclid, 1908][4]. Bob then sends Alice N and e, his so-called *public* key, keeping d as his secret, *private* key.

Alice now encodes her message as some number (often in binary digits, or *bits*) x with the only condition that $x < N$. Using Bob's public key, she computes the encrypted message,

$$y = E(x) = x^e \quad \mathrm{mod} \ N, \tag{2.3}$$

and sends this to Bob. To decrypt Alice's message, Bob computes

$$x = D(y) = y^d \quad \mathrm{mod} \ N. \tag{2.4}$$

Why does this work? Since $d \cdot e = 1 \ \mathrm{mod} \ \varphi(N)$, there must exist some integer k such that $e \cdot d = 1 + k \cdot \varphi(N)$. Thus

$$\begin{aligned} D(y) &= x^{e \cdot d} \quad \mathrm{mod} \ N \\ &= x^{1 + k \cdot \varphi(N)} \quad \mathrm{mod} \ N \\ &= x \cdot x^{k \cdot \varphi(N)} \quad \mathrm{mod} \ N \\ &= x \cdot \left(x^{\varphi(N)}\right)^k \quad \mathrm{mod} \ N \\ &= x \quad \mathrm{mod} \ N \\ &= x. \end{aligned} \tag{2.5}$$

The second to last step follows from Euler's theorem, which states that, for any positive, coprime integers n and a, $a^{\varphi(n)} = 1 \ \mathrm{mod} \ n$. Thus the expression $\left(x^{\varphi(N)}\right)^k \ \mathrm{mod} \ N$ is simply equal to $1^k \ \mathrm{mod} \ N = 1 \ \mathrm{mod} \ N$. To justify the trivial final step in the proof, recall the condition $x < N$.

2.1.2   IS RSA SECURE?

We have shown that the decryption function $D(y)$ is indeed the inverse of the encryption function $E(x)$, so Bob is able to recover Alice's message. However, we have not adequately understood why the transmission of the encrypted message is secure. After all, Bob openly publishes some very fundamental information about his key, namely e and N. So why is an eavesdropper unable to use that information to recover x?

---

[3]Two numbers are said to be *coprime* if they do not share any common factors other than 1.
[4]The notation mod N denotes arithmetic *modulo-N*. That is, the equivalence relation $r = s \ \mathrm{mod} \ m$ is the same thing as saying m evenly divides $(r - s)$.

The answer comes from the fact that, since e is a member of the group of integers *modulo-N*, it has a unique inverse, namely d, a fact that we proved in equation (2.5). Because Bob keeps d secret, the eavesdropper (let's call her Eve) has no direct knowledge of how to compute $D(y)$. Furthermore, in order to compute d, Eve must know the value of $\varphi(N)$. It turns out that computing $\varphi(N)$ is no simple task. The most obvious (but least efficient) way would simply be to count the number of primes less than N. However, for values of N that have hundreds or thousands of digits, performing a primality test on every integer between 0 and N becomes computationally arduous. From the properties of the φ function mentioned above, we see that the only efficient way to compute $\varphi(N)$ is to know the prime factorization of N. Thus, in order for Eve to compute d, she must be able to factor N back into $p \cdot q$. For this reason, φ is often called a *trapdoor* function; it is difficult to compute the answer unless one learns a key piece of information that makes it simple - a trap door.

While it seems simple to write down, efficiently factoring N back into $p \cdot q$ turns out to be a fundamentally difficult, if not impossible, problem. In fact, an efficient prime factorization algorithm is a true 'holy grail' of modern mathematics. In terms of computational complexity theory, it is part of a class of problems known as *Nondeterministic Polynomial time - complete*, or NP-complete. NP-complete problems have two important properties. First, there must exist an efficient (i.e., polynomial-time) algorithm for verifying a given solution to a problem (in our example, it is obvious that knowledge of p allows Eve to verify that it is indeed a factor of N by simple division). The second property of NP-complete problems is that no efficient solutions to any of them have been found. That is, there are no polynomial time algorithms for solving any of the NP-complete problems, of which there are over 3000 known to date (including some other famous problems such as the Traveling Salesman and Graph Coloring). One additional interesting property of NP-complete is that, were one to uncover an efficient solution to one problem in NP-complete, that solution would be applicable to *all* NP-complete problems.

To date, the most efficient prime factorization algorithm is the General Number Field Sieve, or GNFS. It is a sub-exponential algorithm with asymptotic running time

$$\mathcal{O}\left(\exp\left(\left(\frac{64}{9}n\right)^{\frac{1}{3}}(\log n)^{\frac{2}{3}}\right)\right). \tag{2.6}$$

Because the running time is longer than polynomial, computational power adequate for computing $N = p \cdot q$ for primes p and q would not be sufficient to factor N back into p and q as they become larger. This asymmetry between multiplying and factoring large numbers allows Alice and Bob to continuously outpace Eve in computational ability, keeping their public-key cryptosystem secure... for now.

2.1.3   FACTORING IN CLASSICAL AND QUANTUM COMPUTATION

While we have said that there are no known efficient solutions to NP-complete problems, we did not say that one could not exist. It is true that solutions have evaded computer scientists and mathematicians for generations, but a definitive proof whether or not a solution exists (the so-called *P=NP* problem) remains one of the most important open questions in mathematics today. By extension, RSA, the most ubiquitous cipher in the world and protector of the world's most sensitive information, remains secure on only an unproven conjecture - namely, that because a solution has evaded generations of computer scientists and mathematicians, one will not be found.

In fact, the vulnerability goes even deeper. There is a significant community of physicists and computer scientists currently studying the concept of quantum computing, a deeper but less developed cousin of quantum encryption that performs computation using bits (called *qubits*) that can exist in a quantum superposition of 0 and 1 at the same time. This profound difference in computational element led Peter Shor to develop a quantum algorithm that can efficiently factor prime numbers using quantum Fourier transforms. I will forgo a discussion of Shor's algorithm, as it requires a significantly deeper treatment of quantum information theory that would probably make this book far too expensive. However, the general gist of the concept is that, because the qubits can exist simultaneously in the 1 and 0 state, one can think of the algorithm as performing all of the steps of factoring at the same time and providing a result when one makes a final measurement. While counter to our common intuition about natural systems, this concept underlies the dramatic improvement in computational ability that the Shor algorithm achieves. Currently, the only catch is that no one has yet figured out how to build a quantum computer large enough to run the algorithm on any meaningful data. The difficulty lies in creating enough simultaneously interacting qubits to perform operations on substantially large numbers. As of this printing, only four qubits have ever been implemented simultaneously in one quantum computer, enabling the system to successfully factor the number 15 [Matthews *et al.*, 2009].

2.1.4   OTHER PUBLIC KEY CRYPTOSYSTEMS

RSA is not the only public key cryptosystem in existence. The Diffie-Hellman protocol [Diffie and Hellman, 1976] is another one that relies not on prime factorization, but a different one way function: the discrete logarithm. In this protocol, Alice and Bob agree on a large prime number p and a base integer g. Alice then chooses a secret integer a and sends Bob her public key, $A = g^a \mod p$. At the same time, Bob chooses his own secret integer b and sends to Alice his public key, $B = g^b \mod p$. Using these public keys and their private integers, Alice and Bob are both able to compute a key $K = B^a \mod p = A^b \mod p = g^{ab} \mod p$. This procedure is secure because the ability to compute g^{ab} from only g^a or g^b, known as the *discrete logarithm*, is just as difficult computationally as prime factorization. In fact, algorithms to solve one are usually adaptable to the other. By using this alternative trapdoor function, Diffie-Hellman key exchange achieves the same level of security as RSA.

It is often said that, should someone succeed in building a quantum computer, we would simply switch to Diffie-Hellman key exchange and all would be secure. However, Peter Shor extended his original factoring result to show that his same quantum algorithm could be used to solve a discrete logarithm in polynomial time as well [Shor, 1994]. In fact, his algorithm would also apply to even more advanced ciphers such as elliptic curve cryptography [Blake *et al.*, 1999], which are based on similar functions to the discrete logarithm.

2.2   INFORMATION THEORY AND SYMMETRIC KEY CRYPTOSYSTEMS

2.2.1   DEFINING "PERFECT" SECURITY

Since we have called into question the security assumptions surrounding much of today's data encryption, perhaps, it is an apt time to consider, as the Monty Pythons would say, "something a bit more interesting." If public key cryptosystems are secure only up to an assumption about an attacker's computational capabilities, is it possible to create a cipher that achieves truly perfect security independent of an attacker's technical prowess?

Claude Shannon set out to answer just that question when he developed the statistical theory of information in the late 1940's[5]. In the process of applying statistical analysis to information and communication, Shannon developed a definition of perfect security that is profound yet simple [Shannon and Weaver, 1949]. Consider a set of possible messages $M_1, \ldots, M_n$, with independent, *a priori* probabilities $P(M_1), \ldots, P(M_n)$. Suppose we have some transformation T that encrypts these messages into $E_1, \ldots, E_n$, where $E_j = T \cdot M_j$. An eavesdropper can intercept some particular E_j and compute the *a posteriori* probability, $P_{E_j}(M_j)$, that E_j resulted from applying T to M_j, the mathematical description of breaking the cipher. *Perfect security* can then be defined as the condition that $P_{E_j}(M_j) = P(M_j)$ for all $j = 1, \ldots, n$. This is equivalent to saying that E_j provides no information about the message M_j. Essentially, the eavesdropper would have to know the contents of the message in order to break the code, rendering pointless the very act of breaking the code in the first place.

What are the conditions on E that can provide such security? If we consider $P(M_j)$ to be the marginal probability of M_j, $P(E_j)$ to be the marginal probability of E_j, and $P_{M_j}(E_j)$ to be the conditional probability of the ciphertext E_j given the message M_j, then we can use Bayes' theorem to compute the complimentary conditional probability,

$$P(E_j|M_j) = \frac{P(M_j) \cdot P_{M_j}(E_j)}{P(E_j)}. \tag{2.7}$$

If we ignore the trivial case where $P(M_j) = 0$, we see that the only way to achieve $P_{E_j}(M_j) = P(M_j)$ is to satisfy the condition that $P_{M_j}(E_j) = P(E_j)$ for all $j = 1, \ldots, n$. In words, this relationship says that each encrypted message could have resulted from applying the cipher transformation to *any* possible mesage. This is the mathematical way of requiring that there must be the same number of unique keys to the cipher as there are messages, and all keys must be equally likely. Only if this necessary and sufficient condition is met, will the cipher achieve truly perfect secrecy.

---

[5]Unlike many pioneering texts in other subjects, Shannon's seminal work on information theory is very readable and illuminating and is strongly recommended to the curious reader. It is also refreshingly brief.

2.2.2 THE VERNAM CIPHER

Throughout the thousands of years that civilizations have been keeping secrets, only one cipher stands out as meeting Shannon's strict requirements. As mentioned in the previous chapter, the cipher itself is as simple as one can imagine. For a message of length N digits, construct a key of N truly random digits and simply add the message and the key digit by digit to create the ciphertext. To decrypt, subtract the same key digit from each ciphertext digit to recover the original message. Because this scheme satisfies Shannon's security requirements, it achieves perfect security.

Of course, this scheme begs the question of how to transmit the key from one side to the other without someone intercepting it. In practice, this problem is just as hard as securely transmitting the original message in the first place. Thus we have arrived at a conundrum. In order to achieve perfect secrecy, we have created a seemingly impossible implementation problem for ourselves.

2.2.3 PSEUDO-RANDOM NUMBER GENERATORS

Beyond the simple matter of key distribution, there are other technical challenges in implementing the Vernam cipher that make it impractical for real-world applications. Recall the condition that the key contain a number of truly random bits as long as the message itself. This implies that the sender must have a source that generates truly random numbers at the same rate at which the message bits are transmitted. Unfortunately, fast sources of truly random numbers have been yet another elusive component of modern cryptography. Many applications employ generators that deliver numbers that are only approximately random and are thus called *pseudo-random* number generators, or PRNGs. These sources often trade speed for randomness (and hence improved security), and vice versa, depending on the application. For example, one of the simplest PRNGs is the linear congruence generator. It operates as follows: Let a, b, and m be fixed, positive integers such that $m > \max\{a, b\}$. Given an initial *seed* value of x_0, the sequence $x_{n+1} = (ax_n + b) \mod m$ is approximately random, meaning that the values of x_n are seemingly unpredictable as long as a, b, and m remain unknown, and assuming m and b are coprime. However, like many such simple PRNGs, the linear congruence generator is easily broken after a relatively short number of observations of the output [Boyar, 1989]. Hence it is not suitable for any applications, including cryptography, beyond trivial ones such as computer games or ones with extreme memory limitations such as embedded processing applications.

A more effective PNRG is the Mersenne Twister. Invented in 1997 by two researchers from Keio University in Japan, its core computation is a bit more involved but results in a period of $2^{19937} - 1$ in its original form. The name is derived from the fact that $2^{19937} - 1$ is a Mersenne prime[6], or a prime number that has the form $2^n - 1$ for some integer n. The Mersenne Twister is computationally fast and results in high quality pseudo-random numbers. Hence it is often employed in Monte Carlo simulations, where fast, high-quality pseudo-random numbers are required. However, the Mersenne

---

[6]Mersenne primes are notoriously difficult to find. Because they are close to powers of 2, they are very large, and new ones are only larger. The grid computing project GIMPS has been wonderfully successful in finding new Mersenne primes, with three new ones discovered in the last two years alone.

Twister is considered insecure for cryptographic applications because it becomes possible to predict its output after only a relatively short number of observations (624 in the original version).

Currently, there are very few methods of generating cryptographically secure pseudo-random numbers. Many are either closed-source, proprietary algorithms that lack adequate verification or are based on encrypting small numbers of bits using some symmetric cryptosystem. These methods only maintain the level of security of the various ciphers that they use and so are equally as vulnerable to attack. Currently, the only algorithm that generates pseudo-random numbers with theoretical cryptographic security is the one of Blum, Blum, and Shub, often abbreviated BBS. Given a seed bit x_0, BBS computes a sequence based on the recursion relation $x_{n+1} = x_n^2 \mod M$, where $M = pq$ is the product of two large prime numbers. Like RSA, BBS is only unpredictable as long as one cannot factor M back into p and q. Thus it is only secure as long as factoring large composite numbers remains difficult.

Other methods include using hardware components or other system features that have intrinsically high entropy. For example, Microsoft's proprietary cryptographic RNG combines various parameters such as the process ID number, the system time, and various performance counters into a hash algorithm to provide sufficiently random numbers. In UNIX-based systems, the file /dev/random uses various system parameters such as keyboard or mouse inputs, temperature measurements, and others to generate pseudo-random numbers with higher entropy. While a good source of pseudo-random numbers during periods of intense system activity, /dev/random (and its Linux cousin, /dev/urandom) can degrade in quality during periods of system idleness.

One project, LavaRnd, employs a hardware source of chaotic bits (originally a LAVA LITE lamp, but currently a CCD chip in a black box) to generate pseudo-random numbers with even higher entropy. Even more recent approaches have employed the amplitude noise of chaotic lasers to generate pseudorandom bits at very high generation rates [Uchida et al., 2008][Reidler et al., 2009]. However, these sources are chaotic rather than truly random and thus remain theoretically deterministic (though practically still very difficult to predict).

In general, this type of encryption is only as good as the random numbers that are behind it, and often poor quality random numbers can lead even good ciphers into insecurity. The case of RC4 (Rivest's Cipher Number 4) is a good example. RC4 is essentially a one-time-pad but with a key that is no longer a truly random sequence. Instead, it is replaced by a pseudo-random key scheduling algorithm. In 2001, Fluhrer, Mantin and Shamir showed that the initial bits of an RC4 encrypted transmission leak significant information about the scheduling algorithm and render the key vulnerable to attack [Fluhrer et al., 2001]. RC4 was employed by a number of wireless internet encryption protocols at the time, many of which have been deemed insecure by the discovery.

We see that computationally based pseudo-random number generators present a key vulnerability in any cryptosystem. As projects such as LavaRnd demonstrate, software-only solutions have intrinsic weaknesses; even the most secure number-theoretic approaches have their potential security vulnerabilities. LavaRnd attempts to go beyond these limitations by employing external hardware. However, even chaotic, macroscopic physical systems are still deterministic and can be vulnerable to sophisticated physical attacks should an eavesdropper have enough motivation. As we will see, only by employing the randomness intrinsic to quantum mechanics will we be able to achieve truly random number generation.

2.2.4 OTHER SYMMETRIC KEY CRYPTOSYSTEMS

In this chapter, we have examined only one symmetric alternative to public key cryptography. This is misleading, as most modern cryptography only employs slower public-key ciphers in order to exchange keys for symmetric cryptosystems. For many years, the standard symmetric cipher was the Data Encryption Standard, or DES, a block cipher selected as the national standard for digital encryption by the National Bureau of Standards in 1977 [National Bureau of Standards, 1977]. The National Security Agency was involved in the original design of DES, and for many years, DES remained the ubiquitous method of encrypting data around the world. However, in 1997, Rocke Verser, Justin Dolske, and Matt Curtin developed a distributed computing system to challenge DES. With the assistance of hundreds of volunteers donating background CPU cycles, the group managed to publicly crack a DES message for the first time. In the years following, others managed to demonstrate similar results, eventually breaking a DES key in under 24 hours [Electronic Frontier Foundation, 1999].

Initially, this led the information security establishment to use Triple DES encryption, simply encrypting the message with DES, then encrypting that ciphertext again with another round of DES, then finally running a third round of DES encryption resulting in a more secure final ciphertext. This stopgap measure was employed until the final design of the Advanced Encryption Standard (AES) was completed, based on the Rijndael cipher of Joan Daemen and Vincent Rijmen. AES is currently the standard for data encryption in the United States, as established by the National Institute of Standards and Technology.

Many feel that AES is sufficiently secure, even in the eventuality that quantum computers make their way into the landscape of technology. However, assumptions about the security of AES are mainly anecdotal, not based in any sound theory like that of the Vernam cipher. Additionally, AES and other symmetric-key cryptosystems are still vulnerable to quantum computational attacks. While they don't use prime factors or discrete logarithms to maintain security, they are vulnerable to key search attacks, where an eavesdropper simply searches for the right key in order to break the encryption. While an obviously difficult problem on a classical computer, key searching turns out to be tractable on a quantum computer, thanks to the work of Lov Grover. Classical computers can search a key space in at best linear time. However, Grover's quantum search algorithm [Grover, 1996] is able to search a key space in $\mathcal{O}(n^{\frac{1}{2}})$, a quadratic speedup. In this way, even symmetric key ciphers are subject to attack by quantum computer should a large enough one come into existence.

2.3   THE FUTURE OF MODERN CRYPTOGRAPHY

As we have seen in the above sections, the cat and mouse game of cryptography continues to this day. As often as the cryptographers develop new ways to keep secrets, cryptologists find more and more ways of stealing those secrets. Whether by employing more clever techniques or increasingly powerful technologies, the current trajectory suggests that the game will continue barring a paradigm shift in the approach to securing data. As the remainder of this book will demonstrate, that paradigm shift can potentially come in the form of quantum cryptography. Just as quantum computation has the potential to make today's encryption standards vulnerable to attack, quantum encryption has the potential to replace them with techniques based in fundamental physical law rather than assumptions about computational complexity. This shift from a focus on computation to one on the physical transmission medium may provide the paradigm shift that can effectively end the cat and mouse game that has plagued information security for centuries.

CHAPTER 3

The Quantum Mechanics of Photons

Quantum crypography relies on the fundamental quantum mechanics of single photons, so it is imperative that we review those basic quantum principles. Of course, we cannot cover all of introductory quantum mechanics here, so it assumed that the reader has a basic knowledge of linear algebra, Dirac notation, and Hilbert spaces, as well as the central postulates of quantum mechanics.

While there are a number of quantum mechanical operators that we can use to describe single photons, we will begin by discussing their polarization. Not only is this the most commonly exploited quantum property of single photons for cryptography, but its description is generalizable to any two-state property of photons.

3.1 THE POLARIZED ELECTROMAGNETIC PLANE WAVE

We begin with Maxwell's equations describing the propogation of electromagnetic radiation in a vaccum. Presented in their simplest form as coupled space- and time-dependent fields, they can be written as

$$
\begin{aligned}
\nabla^2 \vec{E} &= \mu_0 \epsilon_0 \frac{\partial^2 \vec{E}}{\partial t^2} \\
\nabla^2 \vec{B} &= \mu_0 \epsilon_0 \frac{\partial^2 \vec{B}}{\partial t^2},
\end{aligned}
\tag{3.1}
$$

where ϵ_0 and μ_0 are the electric and magnetic constants [Hecht, 2002]. Recall that one rather general solution to these equations describes a monochromatic plane wave propagating in the z direction, with the electromagnetic field oscillating in the $x - z$ plane and the magnetic field oscillating in the $y - z$ plane, described mathematically as

$$
\begin{aligned}
\vec{E}(z, t) &= \hat{x} \cdot E_0 \exp\left(i\left(kz - \omega t\right)\right) \\
\vec{B}(z, t) &= \hat{y} \cdot B_0 \exp\left(i\left(kz - \omega t\right)\right),
\end{aligned}
\tag{3.2}
$$

where $k = \frac{2\pi}{\lambda}$ is the magnitude of the wavevector and ω is the angular frequency. This wave, depicted in Figure 3.1, we call *horizontally* polarized, referring to the direction of its electric field vectors[1]. A *vertically* polarized plane wave would instead have an electric field described by $\vec{E} = \hat{y} \cdot E_o \exp\left(i\left(kz - \omega t\right)\right)$, with a suitably orthogonal magnetic field.

---

[1] For our purposes, we assume, without any loss of generality, that the word horizontal refers to anything in the $x - z$ plane.

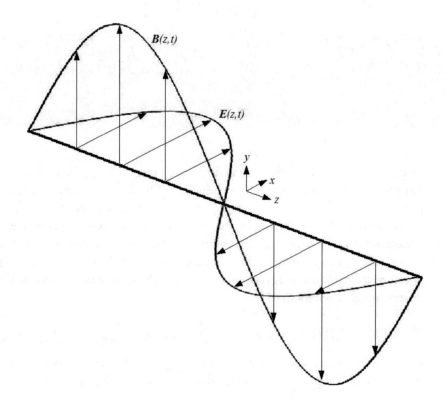

Figure 3.1: *An electromagnetic plane-wave polarized in the x direction.*

Because both x and y polarized plane waves are solutions to Maxwell's equations, any properly normalized superposition of these two polarizations is also a valid solution. Consider, for example, the simple superposition

$$\vec{E}(z, t) = \frac{1}{\sqrt{2}} \left(\hat{x} + \hat{y} \right) E_0 \exp \left(i \left(kz - \omega t \right) \right), \tag{3.3}$$

drawn in Figure 3.2. The resulting electric field points at 45° to the x axis, so we describe this wave as linearly polarized at $+45^o$. If instead the unit vector setting the direction of the electric field were $\frac{1}{\sqrt{2}} \left(\hat{x} - \hat{y} \right)$, the wave would be linearly polarized at -45°. A wave with this polarization is shown in Figure 3.3.

Finally, note that we are describing the electric field in its most general, complex form. This allows us to describe complex phase shifts between the x and y polarized components of the super-position. For example, consider a wave $\vec{E}(z, t) = \frac{1}{\sqrt{2}} \left(\hat{x} + i \hat{y} \right) E_0 \exp \left(i \left(kz - \omega t \right) \right)$. Visually, this is

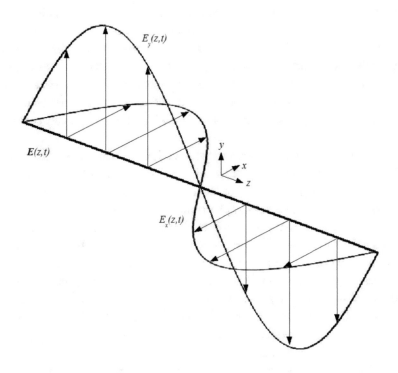

Figure 3.2: *The electric field components of an electromagnetic wave polarized at $+45^o$, resulting from the superposition of waves polarized in the x and y directions.*

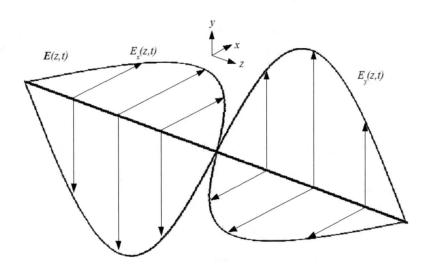

Figure 3.3: *The electric field components of an electromagnetic wave polarized at -45°.*

equivalent to drawing the y polarized component with a $\pi/2$ phase shift relative to the x polarized component, as shown in Figure 3.4. What is the final orientation of the electric field in this case? Suppose we were to view the electric field at various intervals looking directly down the z axis. Different snapshots in time are shown in Figure 3.5. Over time, the resulting electric field vector traces out a circle in the $x - y$ plane. Thus we refer to a wave of this type as *circularly polarized*. The handedness of circularly polarized light is determined by the sign of the y coefficient in the complex superposition $\hat{x} \pm i\hat{y}$, with right-hand circularly polarized light conventionally corresponding to the positive case.

In general, we can describe the various polarizations of plane waves as a two dimensional complex Hilbert space, using column vectors to describe the state of polarization and 2×2 complex matrices called *Jones matrices* [Jones, 1941] to describe various changes to that polarization state in a given physical system. One of the best examples is the use of a waveplate. Used extensively in optics, waveplates are optical elements that have a different index of refraction (and thus propagation speed) along orthogonal polarization axes. This property, called *birefringence*, allows one to change the state of polarization of an incident electromagnetic wave by varying the amount of material through which the wave propagates. For example, a half-wave plate is a birefringent optical element whose thickness is designed to retard the phase of one component (along the "slow" axis) of the electric field by exactly π relative to the field on the other axis (called the "fast" axis). How does this affect the polarization? Suppose a wave linearly polarized at $+45^o$ is incident on a properly oriented half-waveplate. Assuming the fast axis of the optical element is aligned with the x axis and the slow axis with the y axis of the EM wave, we see that the initial polarization vector $\frac{1}{\sqrt{2}}\left(\hat{x} + \hat{y}\right)$ is now replaced with one that contains a π phase shift on the y component, namely $\frac{1}{\sqrt{2}}\left(\hat{x} + e^{i\pi}\hat{y}\right) = \frac{1}{\sqrt{2}}\left(\hat{x} - \hat{y}\right)$. The new EM wave, shown in Figure 3.3, is still linearly polarized, albeit now at -45° instead of $+45^o$. We see that the action of a properly oriented half-waveplate is to rotate a linear polarization by 90°.

Using Jones calculus, we can write the states of polarization before and after the half-waveplate as two dimensional complex vectors. By convention, we will use an orthonormal basis where x and y linearly polarized waves are described as

$$\hat{\epsilon}_x = \begin{pmatrix} 1 \\ 0 \end{pmatrix}, \hat{\epsilon}_y = \begin{pmatrix} 0 \\ 1 \end{pmatrix}. \tag{3.4}$$

Then $+45^o$ linearly polarized light is trivially described by the Jones vector $\hat{\epsilon}_{+45^o} = \frac{1}{\sqrt{2}}\begin{pmatrix} 1 \\ 1 \end{pmatrix}$. The most general form of a Jones vector is one describing an arbitrary complex phase shift θ on one component of the electric field, given by

$$\hat{\epsilon}_e = \begin{pmatrix} 1 \\ e^{i\theta} \end{pmatrix}. \tag{3.5}$$

Physically, what does a field polarized in this state look like? If we observe the total polarization vector from a perspective directly on the z axis as before, we see from Figure 3.6 that the field traces

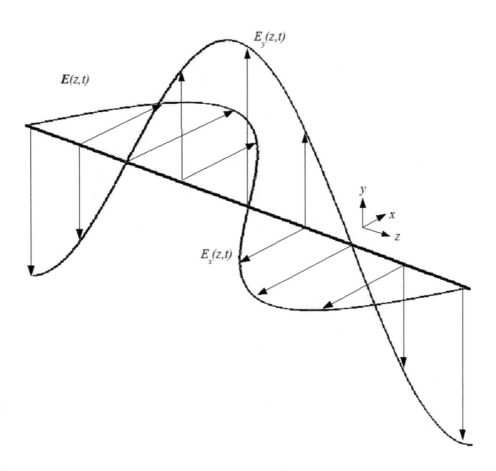

Figure 3.4: *The electric field components of a right-hand circularly polarized electromagnetic wave.*

Figure 3.5: *Snapshots in time (or space) of the total electric field polarization when the y component is shifted in phase by one quarter of the wavelength. The vector traces out a circle in the x-y plane and is called circular polarization.*

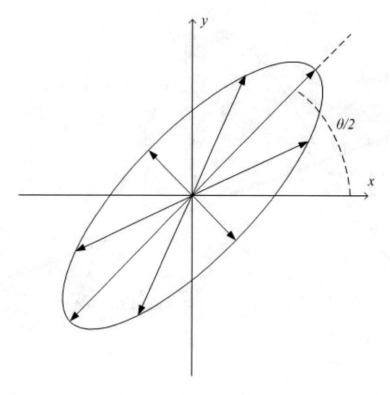

Figure 3.6: *Snapshots in time (or space) of the field vector of an elliptically polarized wave trace out an ellipse in the x-y plane whose eccentricity and orientation are a function of the phase parameter θ.*

out an ellipse with eccentricity and orientation dependent on the phase θ. Because of the generality of this representation, we can derive a continuous range of elliptical polarization states, including the special case of linear polarization, by simply varying the parameter θ, as shown in Figure 3.7.

The Jones matrix describing a half-waveplate is one that transforms the vector $\hat{\epsilon}_{+45^o}$ to the vector $\hat{\epsilon}_{-45^o} = \frac{1}{\sqrt{2}} \begin{pmatrix} 1 \\ -1 \end{pmatrix}$. We can see that this matrix is given by

$$J_{HWP} = \begin{pmatrix} 1 & 0 \\ 0 & -1 \end{pmatrix}. \tag{3.6}$$

Other Jones vectors describing various states of polarization are listed in Table 3.1, while the Jones matrices describing various optical elements are listed in Table 3.2. One interesting aspect of the Jones matrices in Table 3.2 is the distinction in unitarity between polarizers and waveplates. This is because polarizers act as projection operators onto one polarization axis, while waveplates correspond to rotations in the vector space. The physical interpretation of the implications lies in the loss of

State of Polarization	Jones Vector
Linearly Polarized along x axis	$\begin{pmatrix} 1 \\ 0 \end{pmatrix}$
$+45^o$ Linear Polarization	$\frac{1}{\sqrt{2}} \begin{pmatrix} 1 \\ 1 \end{pmatrix}$
-45° Linear Polarization	$\frac{1}{\sqrt{2}} \begin{pmatrix} 1 \\ -1 \end{pmatrix}$
Right-hand Circular Polarization	$\frac{1}{\sqrt{2}} \begin{pmatrix} 1 \\ i \end{pmatrix}$
Left-hand Circular Polarization	$\frac{1}{\sqrt{2}} \begin{pmatrix} 1 \\ -i \end{pmatrix}$

Optical Element	Jones Matrix
Linear Polarizer aligned to x axis	$\begin{pmatrix} 1 & 0 \\ 0 & 0 \end{pmatrix}$
Linear Polarizer aligned to y axis	$\begin{pmatrix} 0 & 0 \\ 0 & 1 \end{pmatrix}$
Linear Polarizer at arbitrary angle ϕ	$\begin{pmatrix} \cos^2 \phi & \cos \phi \sin \phi \\ \cos \phi \sin \phi & \sin^2 \phi \end{pmatrix}$
Quarter Wave Plate	$\begin{pmatrix} 1 & 0 \\ 0 & -i \end{pmatrix}$
Half Wave Plate	$\begin{pmatrix} 1 & 0 \\ 0 & -1 \end{pmatrix}$

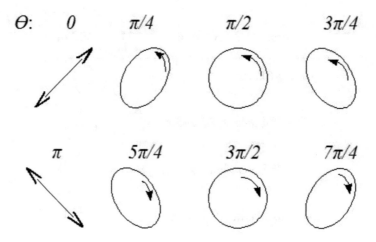

Figure 3.7: *The various elliptical polarization states determined by the phase parameter θ.*

energy or as we will see in the quantum mechanical extension, information about the polarization state. That is, both a $+45^o$ wave and a -45° wave propagating through a linear polarizer oriented along x or y will result in a wave that is polarized along the axis of the polarizer but with a lower intensity than the incident field. Furthermore, phase information that led to the determination of the sign of the original polarization is lost in the projection. That is, there is no way to tell after the polarizer whether the incident light was polarized at $+45^o$ or -45°.

To see the utility of Jones matrices in such an analysis, consider an incident wave polarized at $+45^o$, as in the above example, passing through a polarizer oriented in the y direction. Its field is described by

$$\vec{E}_i(z, t) = \frac{1}{\sqrt{2}} \left(\hat{x} + \hat{y}\right) E_0 \exp\left(i\left(kz - \omega t\right)\right). \tag{3.7}$$

Since the vertical polarizer acts to project the polarization of $\vec{E}_i$ onto the y axis, we can write the field propagating out of the polarizer as

$$\vec{E}_o(z, t) = \frac{1}{\sqrt{2}} \hat{y} E_0 \exp\left(i\left(kz - \omega t\right)\right). \tag{3.8}$$

We know that intensity is related to the square of the field amplitude, so we can calculate the ratio of the output energy to that of the input, given by

$$\frac{I_o}{I_i} = \left(\frac{\frac{1}{\sqrt{2}}}{1}\right)^2 = \frac{1}{2}. \tag{3.9}$$

So we see that the action of the polarizer reduces the field intensity by half[2]. Were we to use Jones calculus to analyze the same situation, we would write the following matrix equation to determine the output state of polarization of the simple system:

$$
\begin{aligned}
\epsilon_o &= \mathbf{J}_y \times \hat{\epsilon}_{+45^o} \\
&= \frac{1}{\sqrt{2}} \begin{pmatrix} 0 & 0 \\ 0 & 1 \end{pmatrix} \begin{pmatrix} 1 \\ 1 \end{pmatrix} \\
&= \frac{1}{\sqrt{2}} \begin{pmatrix} 0 \\ 1 \end{pmatrix}.
\end{aligned} \tag{3.10}
$$

We see that the resulting polarization lies along the y axis, but it does not have unity magnitude. The relative intensity of the output is related to the squared magnitude of the resulting Jones vector. The resulting output intensity is $\left| \frac{1}{\sqrt{2}} \begin{pmatrix} 0 \\ 1 \end{pmatrix} \right|^2 = \frac{1}{2}$, the same as the result in equation (3.9).

3.2   THE QUANTUM MECHANICAL POLARIZATION STATE

In general, any linearly polarized photon at an angle ϕ has the Jones vector

$$
\hat{\epsilon}_\phi = \begin{pmatrix} \cos \phi \\ \sin \phi \end{pmatrix}. \tag{3.11}
$$

Suppose a wave polarized in this state passes through a linear polarizer oriented along the x axis. We know using Jones calculus that the resulting output wave will be polarized along the x axis with intensity

$$
\frac{I_o}{I_i} = \sin^2 \phi. \tag{3.12}
$$

But what happens when the input intensity, I_i, is so small that it effectively contains a *single* photon? We can no longer discuss relative intensities because an electromagnetic wave cannot contain a fraction of a photon[3]. So how do we treat this situation?

We have seen that we can represent the polarization states of classical electromagnetic plane waves by two-dimensional complex vectors. At the quantum mechanical level, the same is true of the polarization states of single photons. While a rigorous treatment of the quantization of the electromagnetic field is beyond the scope of our treatment, it suffices to say that the same calculus that describes polarization in the classical sense works equally well to describe the two-dimensional Hilbert space that contains the polarization states of a single photon. Similarly to the classical treatment, we can choose any two orthogonal states as our basis set, but it is typically most natural

---

[2]Real polarizers either reflect or absorb the remaining energy, leading to potential problems with heating or backreflections that one must take into account in real optical systems.

[3]It is assumed that the reader has a basic knowledge of quantum mechanics and Dirac notation and understands the concept of single quanta of electromagnetic energy. For an excellent introduction see [Griffiths, 2005].

to use the basis set $\{|H\rangle, |V\rangle\}$. The corresponding matrix representation is

$$|H\rangle = \begin{pmatrix} 1 \\ 0 \end{pmatrix}, |V\rangle = \begin{pmatrix} 0 \\ 1 \end{pmatrix}. \tag{3.13}$$

Note that this matrix representation of the quantum mechanical polarization state is exactly the same as the Jones vector. In fact, like the Jones vectors, we can write all of the other polarization states as superpositions of these basis states. Some examples are

$$
\begin{aligned}
|+45\rangle &= \frac{1}{\sqrt{2}}(|H\rangle + |V\rangle) \\
|-45\rangle &= \frac{1}{\sqrt{2}}(|H\rangle - |V\rangle) \\
|R\rangle &= \frac{1}{\sqrt{2}}(|H\rangle + i|V\rangle) \\
|L\rangle &= \frac{1}{\sqrt{2}}(|H\rangle - i|V\rangle),
\end{aligned}
\tag{3.14}
$$

where the $|R\rangle$ and $|L\rangle$ states correspond to a right-hand and left-hand circular polarization, respectively.

Using this basis set, we can also examine the quantum mechanical representations of the same linear optical elements that we considered in the classical case. For example, consider a horizontal polarizer. As we noted above, the action of a polarizer is to *project* a polarization state onto its axis. Hence we can describe a linear polarizer quantum mechanically using a projection operator. Our horizontal polarizer is then

$$
\begin{aligned}
\hat{P}_x &= |H\rangle\langle H| \\
&= \begin{pmatrix} 1 \\ 0 \end{pmatrix} \begin{pmatrix} 1 & 0 \end{pmatrix} \\
&= \begin{pmatrix} 1 & 0 \\ 0 & 0 \end{pmatrix}.
\end{aligned}
\tag{3.15}
$$

This matrix, which is identical to the Jones matrix describing the same element in the classical sense, happens to fit the criteria of an observable - that is, it is Hermitian and has real eigenvalues. Thus inserting a polarizer in front of a single photon corresponds in the quantum mechanical sense to a *measurement* of its polarization state. Suppose before the measurement the photon is in some arbitrary polarization state $|\psi\rangle$. After passing through the polarizer, the photon will be transmitted (the outcome with eigenvalue 1) with probability $\langle\psi|\hat{P}_x|\psi\rangle$, and absorbed or reflected (the outcome with eigenvalue 0) with probability $1 - \langle\psi|\hat{P}_x|\psi\rangle$.

One of the simplest yet most applicable examples would be to let $|\psi\rangle = |+45\rangle$. What outcomes would be possible, and with what probabilities, were we to transmit this single $|+45\rangle$ photon through our horizontal polarizer? We can compute the answer using only the definitions of the states and the fact that our basis set is orthonormal:

$$
\begin{aligned}
\langle +45|\hat{P}_x|+45\rangle &= \frac{1}{\sqrt{2}}\left((\langle H|+\langle V|)\,(|H\rangle\langle H|)\,\frac{1}{\sqrt{2}}\,(|H\rangle+|V\rangle)\right) \\
&= \frac{1}{2}\left((\langle H|H\rangle+\langle V|H\rangle)\cdot(\langle H|H\rangle+\langle H|V\rangle)\right) \\
&= \frac{1}{2}.
\end{aligned}
\tag{3.16}
$$

Obviously, we would get the same result using a vertical polarizer $\hat{P}_y$ instead of a horizontal one. Note that, numerically, the result matches the classical case exactly. The only difference is in the interpretation. Instead of relating the value of $1/2$ to the relative intensity, we now interpret the value as the probability of occurrence of some event related to our single photon. The difference is subtle but profound, and we will see that the implications of such a simple experiment are central to the ideas of quantum cryptography.

There is one more important quantum mechanical property of single photon measurements that we must examine - the concept of simultaneous measurements. We know that we can represent $\hat{P}_x$ by the matrix $\begin{pmatrix} 1 & 0 \\ 0 & 0 \end{pmatrix}$, and we can surmise that a vertical polarizer can be written as $\hat{P}_y = \begin{pmatrix} 0 & 0 \\ 0 & 1 \end{pmatrix}$. Quantum mechanically, are these measurements mutually exclusive? Does measuring one interfere with a measurement of the other? We can answer this question through the use of commutators by the following computation:

$$
\begin{aligned}
[\hat{P}_x,\hat{P}_y] &= \hat{P}_x\hat{P}_y - \hat{P}_y\hat{P}_x \\
&= \begin{pmatrix} 1 & 0 \\ 0 & 0 \end{pmatrix}\begin{pmatrix} 0 & 0 \\ 0 & 1 \end{pmatrix} - \begin{pmatrix} 0 & 0 \\ 0 & 1 \end{pmatrix}\begin{pmatrix} 1 & 0 \\ 0 & 0 \end{pmatrix} \\
&= \begin{pmatrix} 0 & 0 \\ 0 & 0 \end{pmatrix}\begin{pmatrix} 0 & 0 \\ 0 & 0 \end{pmatrix} = 0.
\end{aligned}
\tag{3.17}
$$

Because the commutator[4] of $\hat{P}_x$ and $\hat{P}_y$ is zero, we know that we can make simultaneous measurements of both the horizontal and vertical polarizations. In fact, this is precisely the action of a polarizing beamsplitter (often abbreviated PBS), which transmits horizontal photons on one path and reflects vertical photons onto another path (usually at 90° to the horizontal path). However, instead of writing the action of a PBS as two separate but simultaneous measurements, we can describe it using the measurement operator

$$
\hat{P}_{xy} = \begin{pmatrix} 1 & 0 \\ 0 & -1 \end{pmatrix},
\tag{3.18}
$$

---

[4]The *commutator* of operators A and B is denoted as $[A, B]$ and is defined as $AB - BA$. If two operators commute, then their commutator is zero. One can show that observable operators that commute can be measured simultaneously without any loss of information.

where an outcome of 1 corresponds to a transmitted $|H\rangle$ photon and an outcome of -1 corresponds to a reflected $|V\rangle$ photon. Now we can very simply examine what happens when we transmit a $|+45\rangle$ photon through a PBS:

$$\left|\langle H|\hat{P}_{xy}|+45\rangle\right|^2 = \frac{1}{2}, \left|\langle V|\hat{P}_{xy}|+45\rangle\right|^2 = \frac{1}{2}. \tag{3.19}$$

Physically, we see that the photon has a 50/50 chance of following the horizontal or vertical path[5]. While this seems obvious, such an operation lies at the heart of quantum cryptography.

3.3   QUANTUM ENTANGLEMENT AND THE BELL INEQUALITY

Entanglement is perhaps the deepest yet most perplexing aspect of quantum mechanics known today. It is a phenomenon that logically follows from quantum theory, has been demonstrated experimentally numerous times, yet remains distinctly controversial precisely because it is so counter to our natural intuition about nature. While it would be impossible to provide a complete treatment of quantum entanglement here, we will attempt to cover the aspects that are most relevant to quantum communications. It is fortuitous that these concepts also happen to be the most profound and fundamental ideas underlying entanglement, so the subject provides a good basis for understanding the concept as it will pertain to future, more involved quantum information applications.

Perhaps the best way to introduce entanglement is to present the thought experiment (often called a *gedanken* experiment) that led to the famous Einstein-Podolsky-Rosen paradox [Einstein *et al.*, 1935]. A good description was originally provided by J. S. Bell in [Bell, 1966], to whom we will return shortly. Bell's argument goes as follows: Consider two spin one-half particles created in the singlet spin state propagating in opposite directions - say, two electrons formed from a pion decay. Suppose two observers, say Alice and Bob, are arbitrarily far apart and attempt to measure the spin of those particles (for example, using a Stern-Gerlach type system[6]). Before they make their measurements, Alice and Bob both know that the electrons must have opposite spin by the simple law of momentum conservation. However, which electron has spin +1/2 and which one has spin -1/2 is completely indeterminate. In fact, both particles are in a superposition state of having both positive and negative spin momentum. Thus if Alice measures the spin of her incoming electron, she will obtain random results, with 50% of her measurements indicating a spin of +1/2 and 50% of her measurements indicting a spin of -1/2.

Here is the key: Bob's measurements, though also completely random, are *perfectly correlated with Alice's!* That is, whatever Alice measures, Bob will observe the opposite spin value. Even though

---

[5] I have foregone an explanation of Dirac's bra-ket notation. For a review of the topic, see Griffiths [2005] or Claude Cohen-Tannoudji and Laloë [1973].

[6] It is assumed the reader has knowledge of the Stern-Gerlach experiment [Gerlach and Stern, 1922], arguably one of the most important physics experiments performed in the last century. For a good primer on the subject of spin, see [Griffiths, 2005]. For a more philosophical and thought-provoking examination, see [Ohanian, 1986].

both electrons were in superposition states when they left the pion, measuring one electron immediately determines the outcome of the other remote measurement. The counterinuitive part that still sparks heated debate among the physics community is that Alice and Bob can be arbitrarily far apart, and no information needs to be communicated between them.

So how does Bob's electron know the outcome of Alice's measurement without some kind of superluminal communication prohibited by the other laws of physics? This is precisely the question that inspired Einstein to coin the phrase "spooky action at a distance," and it is the concept that made him most uncomfortable with quantum mechanics as a whole. However, experiment after experiment has verified that indeed the action is real, and each one has discounted another 'loophole' around the fundamental action. For example, experiments have disproved the existence of hidden variables that determine the electron's spin prior to the measurement but cannot be observed [Aspect et al., 1981]. Others have gone further, demonstrating that local realism, the now defunct idea that physical properties are ultimately local to their owners and are fundamentally deterministic, is not correct. The underlying intuitive mechanism still eludes us. It may be that as humans, we simply have no intuition about such matters and must rely solely on the logic and observation of scientific inquiry. However, the philosophical implications are equally fascinating; fundamentally, one can no longer consider the two electrons in the above gedanken experiment to be independent objects. Instead, one must think of them as a single quantum object, a super-particle if you will, that is arbitrarily large. Since wavefunctions are defined everywhere in the universe, we can think of the entangled pair of electrons as a single wavefunction upon which both Alice and Bob act. This is the sobering concept of nonlocality: Everything exists everywhere, at least technically speaking.

Whatever the philosophical implications, the practical implications to information theory are equally fascinating. Later results have shown that, even though there are immediate correlations among Alice's and Bob's measurement outcomes, no actual information can be transmitted across such a link. This result makes sense since the gedanken experiment we just described still results in random, albeit correlated, outcomes for both Alice and Bob. These random results contain no information, implying no violation of the relativistic prohibition on superluminal information transfer. However, if we restrict our interests to simply agreeing on random sequences rather than actually transmitting information, we find that entanglement, especially when applied to photons rather than electrons, becomes profoundly useful.

One last important result pertaining to entanglement is Bell's inequality. Intended as a re-statement of the EPR thought experiment, Bell's inequality provides us with a highly useful metric of entanglement. Consider the shown in Figure 3.8. Suppose a source of entangled photon pairs sends half of each generated pair to Alice and Bob, respectively, who each measure their polarization independently using a polarizing beamsplitter and their '+' and '-' single-photon detectors. Now suppose Alice and Bob can each rotate their beamsplitters through one of four angles, 0, $\frac{\pi}{8}$, $\frac{\pi}{4}$, and $\frac{3\pi}{8}$, and then jointly measure the number of coincidence counts, N_{xy}, defined as the number of times Alice's x detector fires simultaneously with Bob's y detector (where x and y can take the values of +

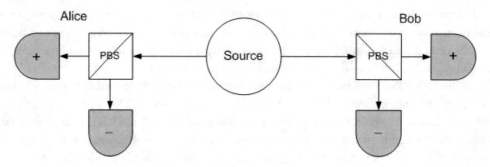

Figure 3.8: *The experiment to test the CHSH form of the Bell Inequality.*

and -, respectively)[7]. We then define the value

$$E(\alpha, \beta) = \frac{N_{++} + N_{--} - N_{+-} - N_{-+}}{N_{++} + N_{--} + N_{+-} + N_{-+}}, \qquad (3.20)$$

where α and β are Alice and Bob's respective beamsplitter angles. Alice and Bob compute the following value after performing the experiment for four different angles combinations:

$$S = \left| E(0, \frac{\pi}{4}) - E(0, \frac{3\pi}{8} + E(\frac{\pi}{8}, \frac{\pi}{4}) + E(\frac{\pi}{8}, \frac{3\pi}{8}) \right|. \qquad (3.21)$$

The result they obtain will depend on whether the source was truly entangled or whether there was some hidden variable determining their outcomes before they made their measurements. If they are observing true quantum entanglement, they should obtain $S > 2$, and if hidden variables exist, then they should see $S < 2$. This is the so-called CHSH form of the Bell inequality, named after John Clauser, Michael Horne, Abner Shimony and Richard Holt, who put Bell's original statement into this more intuitive form [Clauser *et al.*, 1969]. Numerous experiments have famously demonstrated a physical violation of Bell's inequality (i.e., $S > 2$), leading to the conclusion that hidden variables, and thus local realism, do not actually exist [Aspect *et al.*, 1981]. From a quantum cryptography point of view, we can equate the idea of a hidden variable with *any* possible side channel of information leakage. As we will see, the inequality will allow us to perform a Bell test to determine if a link is truly entangled, or if there exists any side channels of information that we have not yet discovered. Ultimately, this concept of entanglement provides us with the most powerful tool to date in achieving secure communications.

---

[7]As we will see later, this entire process looks strikingly similar to a QKD protocol.

CHAPTER 4

Fundamentals of Quantum Key Distribution

Now that we have reviewed some basics of both cryptography and quantum mechanics, we are ready to delve into how combining the two can achieve the ultimate in secure communications. We will start with the simplest of systems and build from there, finally tackling the most profound QKD concept: entangled photon communication. By the end of this chapter, the reader should have a good understanding of how quantum mechanics can provide a fundamental shift in the cryptography paradigm and how QKD can, in theory, make a claim to unconditional security.

4.1 COMMUNICATING WITH SINGLE PHOTONS

4.1.1 QUANTUM RANDOM NUMBER GENERATION

The quantum mechanics behind quantum cryptography can seem trivial in comparison with more involved subjects such as atomic or molecular physics. However, we will see that, when applied to digital communications using single photons, their implications are profound. The simplest example of such an impact is in the field of random number generation. Recall from Chapter 2 that theoretically sound methods of generating cryptographic random numbers are elusive and limited by the same security threats as the encryption methods they are intended to support. Even the hardware based solutions are not truly random in that they ultimately rely on deterministic processes.

Now imagine a quantum mechanical alternative based on the simple polarizing beamsplitter experiment described in Chapter 3. Suppose one added a single photon detector to each output arm of the polarizing beam splitter and again launched single $+45^o$ polarized photons into the device, as shown in Figure 4.1. Suppose every time the H arm detector 'clicked,' we wrote down a 1, and every time the V arm detector 'clicked,' we wrote down a 0. What would we end up with? Since we would produce 1's and 0's each with 50% probability, we would obtain a truly random bit sequence. In fact, if all of the components were perfectly efficient and aligned, this simple source would produce the most random sequence possible in the natural world! Of course, when implemented with real components, such a system does experience limits in quality and speed, but discussion of these effects is reserved for following chapters.

Another approach of Paul Kwiat, et al., to quantum-mechanical random number generation has much more promise as a practical system and is shown in Figure 4.2 [Wayne *et al.*, 2009]. We will show in the following chapters that the output of a coherent photon source such as a strongly

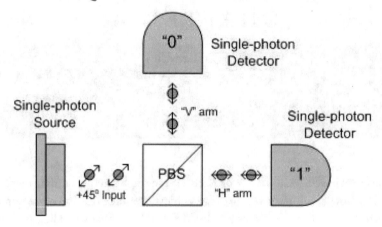

Figure 4.1: *A simple quantum random number generator.*

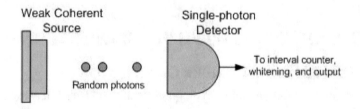

Figure 4.2: *The photon time-of-arrival quantum random number generator. A weak coherent source (such as a diode laser) will continuously generate photons whose arrival times at the detector are exponentially distributed. By timing the intervals between detection events and whitening the data electronically, they can generate very robust quantum random numbers.*

attenuated laser emits pulses whose photon number is Poisson-distributed. This implies that the time between each photon follows an exponential distribution. By using a start-stop counter to digitize the time intervals between each photon detection event from a strongly attenuated laser, and by employing a data whitening scheme, Kwiat and colleagues were able to demonstrate a robust quantum random number generator with real practical potential.

4.1.2   CONJUGATE CODING

One step deeper into the quantum mechanics of single polarized photons is the idea of quantum conjugate coding. While the initial ideas of quantum key distribution were first presented in 1984 by Charles Bennett of IBM Research and Giles Brassard of the University of Montreal [Bennett and Brassard, 1984], the concepts of quantum coding underlying their work originated in

Stephen Wiesner's ideas on conjugate coding [Wiesner, 1983]. At Bennett's behest, the reclusive Wiesner eventually published his work, which subsequently led Bennett and Brassard to develop the initial concept of quantum cryptography.

Wiesner, then a graduate student at Columbia University, originally formulated the concept of conjugate coding with application to transmitting mutually exclusive pairs of messages. The concept is a simple extension of the random number generator. Suppose one took the beamsplitter from the previous section and added in front of it a properly oriented quarter waveplate that was removable. Now suppose one wanted to use this as a receiver to decode messages sent in one of two bases. The two bases considered are $\{|H\rangle, |V\rangle\}$ and $\{|L\rangle, |R\rangle\}$, where the second basis consists of left- and right-hand circular polarization states, and the binary values of the messages are coded such that a 0 would correspond to the first state of a given basis and a 1 would correspond to the second state. Wiesner showed that if the quarter waveplate is installed, only messages sent in the $\{|L\rangle, |R\rangle\}$ basis can be received with any fidelity on the receiver; the messages coded in the $\{|H\rangle, |V\rangle\}$ basis would provide no information since each photon would be equally likely to go to either arm of the receiver. Indeed, messages coded in the $\{|H\rangle, |V\rangle\}$ basis would look just like random receiver noise and could be eliminated through error correcting codes contained in the message itself (see Chapter 5). Furthermore, were one to remove the waveplate, the situation would reverse. That is, in this case only messages sent in the $\{|H\rangle, |V\rangle\}$ basis could be decoded, messages sent in the $\{|L\rangle, |R\rangle\}$ would look like random noise. For this reason, Wiesner dubbed these *conjugate coding bases* and went on to prove that in any Hilbert space of dimension $2^{\frac{(N-1)!}{2}}$ has N mutually conjugate coding bases[1].

4.1.3   THE BB84 PROTOCOL

In 1984 Bennett and Brassard extended Wisner's ideas from mutually exclusive messages to the more applicable concept of data encryption. Their protocol, named BB84 for obvious reasons, again involves four polarization states organized into two non-orthogonal bases. The procedure is as follows: Suppose someone named Alice has a very sensitive message that she wants to transmit to her colleague Bob in a perfectly secure way. Of course, we have shown that the only cipher that she can choose is a one time pad. But how can she give Bob a random key that is as long as her message without any possibility of it being intercepted? Because Alice is clever, she realizes that she can use Wiesner's ideas of conjugate coding, albeit with one modification, to securely send to Bob a stream of randomly polarized that he can use to reconstruct a random key suitable for a one time pad. Suppose instead of having a removable waveplate in front of the PBS, Alice instead places a *non*-polarizing beamsplitter in the path, randomly sending each transmitted single photon to one arm or the other. On one arm, she permanently installs the waveplate and installs a PBS with two single-photon receivers, much like Wiesner's original receiver. On the other arm, she simply installs

---

[1]We will not concern ourselves with the proof here, only with the fact that for the 2D Hilbert Space describing photon polarization, there are three mutually conjugate coding bases $\{|H\rangle, |V\rangle\}$, $\{|L\rangle, |R\rangle\}$, and $\{|+45\rangle, |-45\rangle\}$. This is consistent with Wiesner's statement since $2^{\frac{(3-1)!}{2}} = 2i$.

a duplicate receiver but without the waveplate. The full setup, which she of course gives to Bob in full view of any potential eavesdroppers, is shown in Figure 4.3.

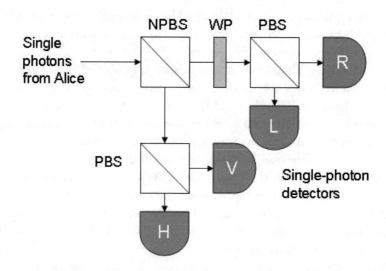

Figure 4.3: *A BB84 receiver. The first nonpolarizing beamsplitter (NPBS) makes the random basis choice, sending each photon to one of two polarizing beamsplitters (PBS), which discriminates between the polarization states. The bases are determined by the orientation of the waveplate (WP).*

Now that Bob has in his possession what Alice has named a BB84 receiver, Alice now considers how she will code her bits. She decides to randomly choose between two of Wiesner's conjugate bases and then to randomly choose which polarization state between the two that comprise her chosen basis to send to Bob. Thus Alice prepares a sequence of single photons randomly polarized in one of four possible states: $|H\rangle$, $|V\rangle$ $|L\rangle$, or $|R\rangle$. She then sends these photons to Bob one at a time over some non-depolarizing optical link. Each time one of Bob's four detectors clicks, he asks Alice over an open communication channel in what basis she chose to send that photon. If it matches the basis that contained his clicking detector, he keeps whatever bit value that corresponds to his detector as the next bit of the random key and tells Alice to do the same, *without actually revealing the value of that bit* (recall that Alice made the original bit choice in the first place). If the bases do not match, they simply ignore Bob's detection, and they repeat the process again and again until they have accumulated enough bits to match the length of Alice's original super-sensitive measurement. This process, called *sifting*, results in a string of random bits that, in a perfect transmission medium with efficient components, should match on both sides. Finally, Alice uses her key to encrypt her message via a one time pad, and Bob uses his copy of the same key to decrypt her message and reveal its contents.

Why is this secure? Consider the theoretical error rate of Bob's sifted key string. In a perfect transmission medium, Bob should always end up with the exact same sequence of random bits as Alice since they ignored any measurements in different bases and thus any introduction of errors. Now suppose an eavesdropper (we will call her Eve) is present between Alice and Bob in an attempt to intercept the key. The only direct attack available to her is a so-called man-in-the-middle (MitM) attack, also called the intercept-resend attack. In this attack, Eve intercepts Alice's single photon transmissions, measures their polarization, and sends new single photons with that polarization to Bob. The question is, can she perfectly recreate Alice's original basis and bit values? From Wiesner's work, we know that the answer is no; she does not know until Bob receives his photon what basis Alice originally used to send. When she tries, she has a 50% probability of choosing the wrong basis. If she does indeed end up choosing the wrong basis, she then has a 50% probability of re-transmitting the wrong bit value to Bob as well. If Bob ends up using the correct basis (with 50% probability), i.e., the same one that Alice used in her original bit transmission, he then has some finite chance of detecting the wrong bit value in the correct basis. Thus the presence of an eavesdropper performing a MitM attack with perfect knowledge of the system will introduce on average a $50\% \times 50\% \times 50\% = 12.5\%$ error into Bob's key values compared the ones that Alice originally sent. If Alice and Bob make the strong assumption that all differences in their respective keys are due to the presence of an eavesdropper, they can use classical error correcting techniques to compare various parameters over the open communcations channel previously used for sifting to effectively eliminate any key bits that may have been comprimised. Once they agree on a sequence of bits, they are sure is known only to them, they are able to use a one time pad cipher to achieve truly unbreakable data encryption.

This seemingly incredible result is supported by two fundamental priciples of quantum mechanics: First, the fact that single quanta, e.g., photons, cannot be split. This comes from the fundamental postulates of quantum mechanics and essentially implies that, once measured, the photon must go to one or the other arm of a beamsplitter[2]. The second result, attributable to William Wooters and Wojciech Zurek, states that a single quantum of unknown state cannot be cloned [Wootters and Zurek, 1982]. This so-called *no cloning theorem* lies at the heart of why Eve is unable to completely recreate the unknown quantum state for the MitM attack. The reasoning behind this limitation traces back to Heisenberg's uncertainty principle. Normally applied to more intuitive conjugate variables like position and momentum (the oft-cited example), Heisenberg uncertainty applies equally to any other conjugate variables, including to measurements made in Wiesner's conjugate coding bases. That is, were one to measure the state of polarization of a single photon in one basis perfectly, it would completely destroy any information about the photons polarization in the conjugate basis, giving each possible value with 50% probability. Thus, unless Eve has additional copies of the quantum state already in hand (which she couldn't, of course, because Alice only sent one), she will not be able to tell what polarization state the photon had when sent.

---

[2]Of course, this does not preclude the photon from being in a superposition state of both arms *before* the measurement. Once the measurement collapses the wavefunction, it must indicate one path or the other.

This protocol, initially developed with polarized single photons in mind, lies at the heart of quantum cryptography. There have been many variations, modifications, and improvements to this protocol in the ensuing years, but with BB84, Bennett and Brassard have achieved, at least in theory, what no other encryption protocol has been able to accomplish in the thousands of years since Julius Caesar. Of course, theoretical security is only the first step in the battle. We will see in later chapters what it takes to implement their concept in the real world and what limitations taking their concept to actual hardware poses.

4.1.4   THE B92 PROTOCOL

In 1992, a number of years after the initial announcement of BB84, Bennett introduced a variation on the four-state protocol that only required two states, albeit with a 50% decrease in efficiency of bit transmissions [Bennett, 1992]. Still, this protocol is rather common in the experimental community becuase it requires half of the state preparation hardware and is convenient in applications where bit efficiency is cheaper than state preparation hardware.

Under the B92 protocol, Alice prepares her photon in one of two non-orthogonal states, say $|H\rangle$ and $|+45\rangle$, that represent a 0 and 1 bit, respectively. Bob then uses the same receiver as he would under BB84, but he interprets the results slightly differently. He purposely measures the photon polarizations in the wrong basis and only keeps the bits that he can measure conclusively. For example, suppose Alice transmits a 1 using a $|+45\rangle$ polarized photon. Now suppose Bob measures this in the seemingly correct basis of $\{|+45\rangle, |-45\rangle\}$. There is no way for him to obtain a result of $|-45\rangle$ (unless an eavesdropper is present, of course), and obtaining a result of $|+45\rangle$ would be ambiguous unless Alice announced the bit value that she sent (an obvious security violation). Suppose instead that Bob measured the $|+45\rangle$ photon in the $\{|H\rangle, |V\rangle\}$ basis. He then knows that the only way his measurement can give a $|V\rangle$ result would be if Alice had transmitted a 1. The opposite situation would be true as well, where Bob would only know if Alice sent a 0 by measuring the photon in the $\{|+45\rangle, |-45\rangle\}$ basis.

Of course, Alice and Bob only maintain security if Bob does not ever reveal his measurement basis. Thus they must employ a modified sifting algorithm (sometimes called *erasure*) where Bob only tells Alice which bits to keep and which to ignore.

4.2   COMMUNICATING WITH ENTANGLED PHOTONS

4.2.1   RANDOM BITS AND INFORMATION LEAKAGE

Though we declared that Bennett and Brassard have achieved the holy grail of data encryption, there are some subtleties that may cause us to reconsider. Note that the two protocols described in the previous section require Alice to make at least one or two random choices before she sends each photon to Bob. This is an intrinsic weakness in the system because it relies on a good random number generator, something that we learned in the previous chapter is difficult to make. Of course, they could use the Blum-Blum-Shub algorithm and obtain good cryptographic random numbers whose security is based on the difficulty of factoring prime numbers. But quantum cryptography is important precisely becuase it does *not* rely on the difficulty of factoring composite numbers, so choosing a PRNG based on such a condition seems counter-productive.

Unfortunately, the intial problems with the above protocols go even deeper. Note that we described Eve's interecept-resend attack but made no mention of any other means by which a clever eavesdropper may gain information. While an ideal system would be impervious to any other attacks in the space between Alice and Bob, real implementations can potentially leak information about the secret key though unintended side channels. By attacking those side channels, Eve could indirectly glean information about the key without affecting Bob's measurements.

What side channels can exist? The answer is as varied as one's imagination. Consider as an example Bennett's first implementation of BB84 QKD at IBM Research Labs in 1989 [Bennett *et al.*, 1992]. His original experimental setup used high-voltage Pockel's cells to prepare the polarization states. These devices, when switched, make audible 'clicking' sounds, alerting any listener in the room to a change in bit or basis. While only a toy example, it illustrates a case where Eve can attack some side channel of information in order to gain knowledge of the key without intercepting the actual photons. Other more sophisticated attacks have been demonstrated, including analysis of timing information on the classical communication channel [Lamas-Linares and Kurtsiefer, 2007] and detection of small reemissions in the single photon detectors [Kurtsiefer *et al.*, 2001]. Similar effects, all the result of the limitations of various components in real-world QKD implementations, are discussed at length in later chapters.

4.2.2   THE E91 PROTOCOL AND THE BELL TEST

In 1991, Artur Ekert, then at Oxford University, found a way around all of the loopholes in BB84 (and, subsequently, B92) QKD. His paper, currently the most cited in the field, describes a key distribution protocol based on the gedanken experiment of Einstein, Podolsky, and Rosen [Ekert, 1991]. The basic concept is illustrated in Figure 4.4. The most obvious difference between this approach, deemed E91, and the BB84 or B92 approaches is that Alice no longer generates single photons in one of four states and sends it to Bob. Instead, both Alice and Bob receive one member of a pair of polarization entangled photons from a source in the middle and then perform measurement, sifting, and reconciliation procedures identical to those in BB84/B92, only on their half of an entangled pair rather than on a single photon. Recall from the previous chapter the idea that Alice's and Bob's polarization measurements will be correlated due to the entanglement but still completely random. This is one of the primary advantages of this approach - that Alice no longer needs to make random bit and basis choices. Thus a good random number generator is no longer required and one of the potential loopholes in single-photon QKD is eliminated.

There is another advantage to entanglement-based QKD. Recall from the previous chapter the strong link between entanglement and information. One extension of the work of Bell and Aspect on hidden variables is the idea that if two particles are entangled, there can be no other way for information to leak out about them via any side channels. This is a very deep result when applied

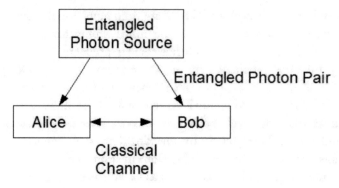

Figure 4.4: *The basic setup for entanglement based QKD using E91.*

to entanglement-based QKD. It implies that if Alice and Bob know for sure that their photons are entangled (something they can measure using a test of the Bell Inequality), they can know *for sure* that no side channels of information can exist in their communications system. This requirement applies equally to channels they may know about or channels that they may never have imagined, and therein lies the depth of the result. By simply performing a Bell test on some of the entangled photons, Alice and Bob can *guarantee* that no information about their secret key is leaking anywhere in the system.

4.3   CODING ON OTHER QUANTUM STATES

The above discussion has focused entirely on the quantum mechanics of and encoding on a photon's polarization. While convenient and intuitive for the purposes of our edification, polarization encoding is only a robust option in free-space optical propagation environments, where the transmission medium does not depolarize the photons as they propagate through it. In other media such as optical fiber, polarization is not a stable parameter and other coding schemes are preferable. Fortunately, the same quantum mechanics outlined above applies equally well to any two-state quantum characteristic of a photon.

More details on these topics will be discussed in later chapters, but it is imperative to illustrate the ideas behind extending the same quantum mechanical algebra to other physical properties of photons. In fact, there is no theoretical reason why the same quantum mechanics does not apply to any two state quantum system (such as electron spins or two-state atomic systems). However, because photons are so fundamental and ubiquitous in communications, we will limit our discussion only to the two-state quantum properties of the electro-magnetic quanta.

4.3.1   PHASE ENCODING

Secure optical communications often evokes the immediate application of optical fibers, and QKD is no exception. Fiber optic quantum key distribution systems are one of the two fundamental implementations that hold promise in bringing quantum cryptography to the marketplace. However, one of the biggest challenges to using fiber optics for QKD is that polarization states are not stable in fiber[3]. Much work has been done to stabilize macroscopic (as opposed to single-photon) polarization states in optical fibers by using asymmetrical cores or by implanting longitudinal strain elements in the fiber cladding. However, no technique developed to date has the fidelity required for long-haul quantum key distribution using polarization states in fiber. Thus one alternative that is often employed is to implement the same BB84 protocol interferometrically, using two conjugate phase states of single photons. The fundamental idea replaces Alice's random choice of polarization with a random choice between two orthogonal photon phase states, often achieved through the use of an electro-optic phase modulator. In a similar fashion to quadrature phase shift keying (QPSK), Alice makes a choice between one of four (or one of two in the case of B92) phase states of the single photon. Bob then employs Mach-Zender interferometry Hecht [2002] between the photon and a reference signal with variable phase delay (for his basis choice) in order to measure the intended phase of Alice's single photon. Of course, this requires Bob to have access to a phase-stable reference signal, a challenge that often adds a layer of complication to phase-encoded QKD systems; we will delve deeper into these challenges in the later chapters.

4.3.2   FREQUENCY ENCODING

Another coding scheme that is more stable in fiber-optic systems than both polarization and phase is one based on modulation sidebands. This encoding scheme uses conjugate sidebands of single photons whose frequency is modulated at the source. Often called Single Sideband (or SSB) coding, this technique does not require a phase-stable reference and so is much more robust in the presence of real-world environments where temperature and strain changes can affect the fiber path length. In addition, optical frequency is much more stable in fiber than is polarization, partly because of the intrinsic nature of the transmission medium and partly because manufacturers have developed much better control over fibers' dispersion characteristics than they have over their polarization control. The specifics of this promising technique will be discussed in later chapters.

4.4   CAVEATS ON THE SECURITY CLAIM

In this chapter, we have begun to understand how to exploit the quantum mechanical nature of single photons to achieve theoretically perfect security. The next chapter discusses the second half of the QKD process - that of error correction and privacy amplification, collectively called *key reconciliation*. This process relies heavily on classical information theory and the use of a reliable public channel of communication, and it is integral to QKD protocols. One obvious implication of this requirement is

---

[3]There are other challenges to fiber-optic QKD, but they will be discussed in later chapters.

that QKD is certainly not immune to a so-called denial of service (or DoS) attack. That is, Eve can easily block all communication by intercepting so many photons that Alice and Bob cannot construct a secure key. The simplest way for Eve to perform this attack would, of course, be to physically block all of the single photons (for example, by simply placing her hand in the beam). However, there are even more subtle attacks that effectively result in a DoS attack, namely anything that results in errors great enough to overwhelm Alice and Bob's reconciliation scheme (described in the next chapter). In fact, even in the absence of an actual eavesdropper, quantum channels that are excessively noisy or error-prone will be inadequate for secure QKD. More on this will be discussed in further chapters, but it is important to note that the phrase unconditional security does imply a few disclaimers. That said, properly implemented QKD will most certainly alert Alice and Bob to their failure to exchange a secure key. The unconditional security claim implies that there is no way for Alice and Bob to be spoofed into thinking they successfully exchanged a secure key when in fact they did not.

Of course, all of this assumes that Eve only has classical (i.e., non-quantum) abilities at her disposal. If one relaxes this assumption, one opens a whole new host of possible attacks on the quantum channel. For example, Howard Brandt has proposed an eavesdropping mechanism based on an entanglement probe [Brandt, 2005]. In this attack, Eve places an entangling system in the path of the quantum channel. This can consist of a nonlinear optical crystal, for example, pumped with a laser that results in output photons that are entangled with Alice's quantum channel single photons. If Eve measures the other half of the entangled pair after Bob performs his measurement on the original single photon, her results will be completely correlated with Bob's, allowing her to recreate the key without being detected. Of course, this attack is purely theoretical – implementing it in real hardware is so prohibitively difficult that it is effectively impossible.

Two final caveats regarding Alice and Bob's secure key exchange: The first is obvious, namely that QKD only claims to secure the space between Alice's transmitting aperture and Bob's receiving aperture. QKD makes no claim when an attacker has access to Alice or Bob themselves, or to the information before it is encrypted or after it is decrypted. Thus should an opponent choose to perform an attack on the nodes of the network and exploit some other human or technical weakness outside of the secure key exchange, the choice of encryption method (QKD or otherwise) will be irrelevant. Secondly, QKD provides no method for authenticating the quantum or classical communications channel prior to the key and data exchange. That is, Eve can preempt Bob in an attack before the communication begins and, assuming she has full knowledge of Bob's hardware, can simply pretend that she is the intended receiver. Thus good QKD requires some other form of authentication, such as the Needham-Schroeder protocol. Good authentication is an entire subject unto itself, so it will be reserved for other books in other sections of the library. Suffice to say, good QKD is useless without secure authentication, so beware of imposters.

CHAPTER 5

Information Theory and Key Reconciliation

By now we see why the phrase "quantum key distribution" is certainly more apt than "quantum cryptography" in describing the processes outlined thus far. In fact, it is an oft misunderstood fact that no actual information is transferred across the quantum channel (in entangled systems, we know it to be impossible) and that QKD depends completely on a parallel classical link for both the error correction/privacy amplification steps as well as the actual secured communications via the quantum key and a one time pad. The design of this link, though not subject to any security requirements, does require authentication and does play an integral role even in the quantum key exchange portion of the communication process.

In the above descriptions of the protocols we mentioned in passing the use of classical error correcting techniques to reconcile the mismatched bits in Alice's and Bob's respective keys. However, there is much subtlety in how this is accomplished in the quantum key exchange process. In fact, as Bennett, Brassard, and Robert point out in their 1988 paper on the subject [Bennett *et al.*, 1988], traditional error correcting codes such as those described in [van Lint, 1982] prove inadequate for the purposes of QKD. This is because traditional error correcting codes assume that errors are sparse and stochastic. However, as we have learned from the above sections, an eavesdropper on the quantum channel does not introduce only a few errors but can create errors in excess of 12.5%. In fact, in an extreme case, the eavesdropper can manipulate the quantum channel in such a way that, though she will never learn the actual value of the bits transmitted, she can induce errors in up to half of the sifted key! No traditional error correcting code is designed to handle such malicious and frequent errors.

Hence in addition to devising the protocols required for the quantum channel key distribution, Bennett and Brassard worked with classical information theorist Jean-Marc Robert to devise a new algorithm for reconciling the mismatched bits in Alice's and Bob's respective bit strings. In essence, they use the essentially perfect but public classical channel to augment the information contained in the imperfect but private quantum channel in such a way that they leave Eve with an arbitrarily small amount of information about Alice and Bob's shared secret. Their approach rests on another breakthrough in classical information theory from two other IBM researchers, J. Lawrence Carter and Mark Wegman, called *universal hashing* [Carter and Wegman, 1979].

5.1   INFORMATION AND ENTROPY

The ultimate goal of the key reconciliation process is to use the public communications channel to estimate how much information an eavesdropper might have about the random bits and to choose the proper subset of those bits in order to minimize her final knowledge of the reconciled key. To complicate matters, Alice and Bob must have this conversation while revealing as little as possible about the bits in which they do have confidence. As we will see, the process is nontrivial. However, before we begin to unravel how they achieve this, we should start by defining exactly what we mean by information.

As we began to learn from the previous chapter's treatment of entanglement, the connection between information and physics is perhaps one of the deepest insights of modern science, and Claude Shannon was perhaps one of the most insightful scientists to understand that connection[1]. He realized that the language used to describe statistical physics was equally applicable to information and that it was inextricably linked to the concept of entropy. Generally speaking, entropy is a measure of the amount of randomness in a system. It seems intuitive then that the less random something is or the more specifically it is defined, the more information is contained in that description, qualitatively linking the concept of information with the concept of randomness.

Mathematically, Shannon defined the *entropy H* of a random variable X with possible values $x_1, \ldots, x_n$ as

$$H(X) = -\sum_{i=1}^{n} p(x_i) \log_2 p(x_i), \tag{5.1}$$

where $p(x_i)$ is the probability that X is equal to x_i. Note that in this definition the logarithm is base 2. This is because we are defining everything in the information units of *bits*, or *binary digits*. Were we to use other logarithmic bases, we would simply be measuring the information content in some other unit (*dits* for base 10, *nits* for base e, etc.).

Another definition of information entropy is actually better suited to our purposes and was originally put forth by Rényi [Rényi, 1961]. It can be considered a higher order of a general definition in which Shannon's entropy is simply the first order and is defined by

$$R(X) = \frac{1}{1-\alpha} \log_2 \left(\sum_{i=1}^{n} p(x_i)^\alpha \right), \tag{5.2}$$

where α is the order of the metric. Specifically, we will see that the Rényi information of order 2, also called the collision entropy, is an important metric in measuring an eavesdropper's information about the quantum key.

---

[1] Another was the late John Wheeler, who believed that information underlies all physical processes and coined the phrase "It from Bit." [Wheeler, 1999]

5.2   UNIVERSAL HASHING FUNCTIONS

Another important information theoretic concept is *hashing*. At the most basic level, hashing functions are functions that take very large data as their input and generate very small data as their
output. In cryptography, we define hash functions as a function that takes a variable sized long input
of bits and produces a fixed, small sequence of bits that is effectively unique. Of course, it cannot be
completely unique but given a set of expected inputs, the probability that two outputs are the same,
also called the *collision probability*, should be small. In addition, good cryptographic hash functions
must be one-way. That is, it must be easy to compute the hash value but difficult to use the hash
value to predict the input. For more on general hashing functions, see [Schneier, 1996].

Different hash functions have different properties, and each one has a set of preferred inputs.
Thus when presented with a set of data that is random and has non-stochastic, indeterminate errors
(such as our quantum bit strings in the presence of an eavesdropper), no single hash function is most
efficient. In cases such as these, universal hashing [Carter and Wegman, 1979] is more appropriate.
Universal hashing is a way of creating a *class* of hash functions that is as efficient as choosing
a hash function completely at random but with better efficiency and more simplicity. A class of
hash functions is universal if, on average, the outputs of its constituent functions are uniformly
distributed, thus minimizing the collision probability. Universal hashing offers another advantage
over simply choosing a hash function at random; in general, defining a hash function at random
requires specifying $K \cdot 2^N$ bits, where K is the number of bits in the hash output and N is the
number of bits in the input. By choosing a hash algorithm from a class of universal functions (also
called universal$_2$ for reasons explained in [Carter and Wegman, 1979]), one can achieve the same
collision probability as a completely random choice while only specifying $2N$ bits. As we will see,
universal hashing provides the necessary security for good quantum key reconciliation.

5.3   PRIVACY AMPLIFICATION VIA THE PUBLIC CHANNEL

Having touched on some of the basic components of information theory, we are prepared to delve into
how these concepts apply to quantum key reconciliation. Suppose X is Alice's randomly generated
bit string of length N that she wishes to share with Bob. Also suppose Eve, the eavesdropper, can
listen to and tamper with X while it is en route to Bob via an eavesdropping function $e : \{0, 1\}^N \rightarrow
\{0, 1\}^K$, $K < N$ and a tampering function $t : \{0, 1\}^N \times \{0, 1\}^M \rightarrow \{0, 1\}^N$. Bob receives a bit string
$Y = t(X)$ that is the result of Eve's tampering. Of course, by the quantum mechanical nature of
the channel, Eve can only know the actual function e and her eavesdropping result $Z = e(X)$. She
cannot physically know $t(X)$ or X itself.

Alice and Bob's overall strategy is to publicly agree on some procedure with which to determine
if $Y = X$. This procedure generally reduces to Alice and Bob collectively choosing a hash function,
f, and comparing their values of $f(X)$ and $f(Y)$. In the limit of breaking down the string into
individual bits, this strategy would allow them to throw out all of the bits that do not agree without
revealing any information about the bits that do agree. Of course, the procedure that they use would

normally be dependent on the nature of the errors present in the string. However, in revealing the nature of the errors and thus the optimal error correcting strategy, they would reveal to Eve all the information she would need to spoof the reconciliation and manipulate the results in her favor.

This is where universal hashing becomes critical to implementing effective key reconciliation. Recall that choosing a hash function from a class of universal hash functions not only avoids revealing the specific function used but also offers an improvement in efficiency over traditional error correction as well, requiring only $2N$ bits of information to describe f rather than $K \cdot 2^N$ required to describe any arbitrary hash function. Thus with universal hashing we have both a decreased collision probability and increased efficiency all while minimizing Eve's ability to manipulate the bits to her advantage.

As mentioned above, it is also important that Alice and Bob choose their function for a given bit *after* that bit is detected. Otherwise, Eve would be able to modify t in order to maximally exploit f. Suppose Alice and Bob do choose some hash $f : \{0, 1\}^N \to \{0, 1\}^K$, $K < N$, $f \in$ universal$_2$. They then compute $f(X)$ and $f(Y)$, respectively, comparing whether $f(X) = f(Y)$ over the classical channel. If not, Bob can use any error correction scheme from simple bit twiddling to a block or convolutional code to correct the errors[2]. Of course, if the errors are excessively frequent, no amount of classical error correction would be able to cope and Alice and Bob would conclude that the bit exchange could not produce a key. However, should Bob be able to correct the errors, then he and Alice would possess a bit string about which Eve has little to no information.

5.4 THE CASCADE ALGORITHM

The most ubiquitous implementation of these principles of quantum key reconciliation uses the CASCADE algorithm of Brassard and Salvail [Brassard and Salvail, 1994]. This algorithm combines the error correction and privacy amplification into a single iterative process that results in Eve knowing less than one bit of information about the quantum key. It can handle error rates as high as 15% and provides security approaching the theoretical limit. The procedure goes as follows:

Suppose Alice and Bob have respective sifted bit strings after a BB84 key exchange represented by $X = x_1, \ldots, x_N$ and $Y = y_1, \ldots, y_N$ where $X, Y \in \{0, 1\}^N$. Alice and Bob agree on some initial parameter k_1, and they both choose $\frac{N}{k_1}$ substrings $X_v^{(1)} = \{x_j | (v - 1)k_1 < j < vk_1\}$ for $v = 1, \ldots, \frac{N}{k_1}$. Alice then sends Bob a parity check on each block $X_v^{(1)}$ and Bob compares that parity to his blocks $Y_v^{(1)}$. He then corrects the errors using an interactive binary search with Alice, comparing the parity of successive sub-blocks of half the length of the previous comparison until all of the errors are corrected. Following this initial error correction step, Alice and Bob choose a new k_i for $i \geq 2$ and randomly choose a universal hash function $f_i : \{0, 1\}^N \to \{0, 1\}^K$ where $K = \frac{N}{k_i}$. Alice constructs a new set of subsequences $X_v^{(i)} = \{x_j | f_i(j) = v\}$ and sends Bob the value $\bar{X}_v^{(i)} = \oplus_j x_j$, the XOR of all of the elements of $X_v^{(i)}$. Bob performs the analogous block construction, computes the XOR of all of the elements in his substrings, and performs another interactive binary search, comparing parities

---

[2]These classical error correction techniques are described in [MacWilliams and Sloane, 1978].

of successively smaller blocks until all of the differences are removed. Once they have completed this procedure enough times, no blocks with odd numbers of errors will remain; eventually, the errors will be fully corrected, and Eve's information about the bit strings will be eliminated. The proof that this algorithm eliminates Eve's information is provided in [Brassard and Salvail, 1994] and need not be duplicated here. Suffice to say Brassard and Salvail illustrate that even with error rates as large as 15%, 4 passes of CASCADE provide security approaching the theoretical limit of less than one bit of information leaked about the key.

5.5   CONCLUSION - THE INFORMATION THEORY ICEBERG

This chapter is meant as an outline of the process of key reconciliation and distillation. A full treatment would require digging much deeper into the topics of noisy coding theory, finite field algebra, and information theory in general. Hopefully, this summary has provided enough detail to make sense of the references, as they are much more thorough and can provide ample detail regarding each of the topics.

One obvious point that this chapter does illustrate is that quantum cryptography as a subject lies at the intersection of a number of disparate disciplines. Expertise in the subject requires knowledge of physics, optics, communications, cryptography, information theory, and, as we will see in the coming chapters, engineering, computer science, and material science as well. For this reason, the topic is simultaneously both deeply challenging and deeply satisfying, so hang in there!

CHAPTER 6

Components for Broadband QKD

Now that we have learned the fundamental theory of quantum key distribution, we can begin to understand how to build systems that implement the QKD protocols using modern hardware. We will first examine a number of the fundamental components with the unique ability to create and detect single photons and then understand how various implementations use these components to create full QKD systems.

A common thread among real QKD systems is the observation that implementing a theoretically secure protocol in real hardware results in systems that only approach this theoretical security bound. We have already seen this effect when looking at pseudorandom number generators or timing side channel attacks. In addition to understanding the hardware used to create QKD links, we will also understand how the limitations in this hardware leads to security loopholes that must be addressed as well.

At their core, single-photon communications systems differ from classical optical communications systems most strongly in their single photon sources and detectors. Between the sources and detectors, the optics are generally similar if not identical to classical optical hardware. However, the devices required to generate and detect single photons can be quite exotic, as we will see. Thus in this chapter, we will focus primarily on the hardware dedicated to this challenging task.

6.1 SEMI-CLASSICAL SOURCES

6.1.1 THE PHOTON NUMBER SPLITTING ATTACK

From an optical engineering point of view, the first requirement of a QKD system is single photons. After all, the entire concept is based on the physics of these electromagnetic quanta. So how do we create single photons?

The most straightforward way would be to take a pulsed classical (or semi-classical) light source such as a laser and simply attenuate the intensity of the beam until it emits only one photon during each pulse. Of course, such sources are not going to emit pure single photon states. In the quantum mechanical description, coherent states[1] are of the form $|\mu e^{i\theta}\rangle$, where μ is the mean photon number of the source and θ is the random phase [Lutkenhaus and Jahma, 2002]. A complete description of a coherent source thus has an ensemble of these states with random phase, emitting

---

[1]For an in-depth description of coherent and Fock states, see [Loudon, 1973]. For a review of pure and mixed states, as well as density matrix notation, see [Claude Cohen-Tannoudji and Laloë, 1973].

the mixed state $\rho = \frac{1}{2\pi} \int |\mu e^{i\theta}\rangle \langle \mu e^{i\theta}| d\theta$. This is equivalent to a mixture of Fock (or pure photon number) states $\sum_n p(n, \mu)|n\rangle \langle n|$, where $p(n, \mu)$ is a Poisson distribution in photon number n, given by

$$p(n, \mu) = \frac{\mu^n \cdot e^{-\mu}}{n!}. \tag{6.1}$$

Thus even if the intensity is turned down so that, on average, only one photon is emitted per pulse, there is a finite chance that zero, two, or even three or more photons will be emitted instead of one. Obviously, the presence of multiple photons in a pulse has a significant security implication; it allows an eavesdropper to potentially split that pulse and measure only one of the photons while leaving the other one (or more) undisturbed, gaining information about the key without being detected. This so-called photon number splitting (or PNS) attack [Lutkenhaus and Jahma, 2002] drives important considerations when using strongly-attenuated semi-classical sources of entangled photons. The most straightforward way to mitigate the effect is to choose a mean photon number such that any information leaked through two or more photon events would be eliminated by the key distillation process described in the previous chapter. Equation (6.1) is plotted in Figure 6.1 for various values of μ. To illustrate the importance of mean photon number on the security of the link, suppose Alice and Bob choose a $\mu = 1$. This would give them a probability of a two-photon event of $p(2, 1) \approx 18.4\%$, a three-photon event $p(3, 1) \approx 6.1\%$, etc. Summing all of the probabilities for events of two or more photons means that Eve would be able to steal over 25% of the bits!

Instead, systems that employ strongly attenuated coherent sources often employ a mean photon number of 0.1 leading to $p(2, 0.1) < 0.5\%$, an acceptably small number and within the assumptions of the key distillation procedures. Of course, $\mu = 0.1$ implies that only one in ten pulses on average contains a single photon while the rest contain nothing (except for the very few that may contain more than one). While Alice and Bob can deal with this simply by sifting bits only when Bob detects a photon, it has significant implications on link efficiency. Specifically, any link using an attenuated coherent source with $\mu = 0.1$ automatically loses a factor of 10 in overall key generation rate due to all of the empty pulses. While mitigating the security effects, these types of systems are far from maximally efficient.

Another approach to mitigating a simple PNS attack is for Alice and Bob to monitor the photon statistics of their transmission. If an eavesdropper is simply stealing photons from the two or more photon events (a difficult but not physically impossible task), then the photon number distribution at Bob's end of the link will be different than the Poisson distribution that Alice observes. However, as Lutkenhaus and Jahma point out in [Lutkenhaus and Jahma, 2002], Eve can mount a more sophisticated attack that preserves the photon statistics, steals photons from the multiphoton events, and simply looks to Alice and Bob like slightly higher overall link loss. Thus either Alice and Bob absorb a loss in link efficiency in order to maintain security, or they must devise a better way to use weak coherent sources for QKD.

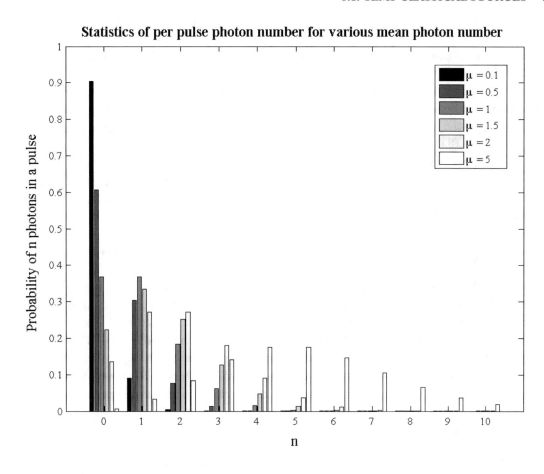

Figure 6.1: *The Poisson distribution of photon number per pulse for various mean photon numbers. Notice that, for a mean photon number of 1, almost 20% of the pulses will contain two photons.*

6.1.2   THE DECOY STATE PROTOCOL

To deal with this loss in key generation efficiency, Won-Young Hwang proposed a protocol that varies the mean photon number in order to probe the link loss and detect the presence of an eavesdropper [Hwang, 2003]. In this modification of BB84, Alice intentionally and randomly inserts multi-photon pulses, referred to as decoy pulses, into the transmission. Because Eve does not know which pulses are real and which are decoys, she attempts to steal photons from all of the pulses. Throughout the transmission, Alice and Bob compare their observed photon number statistics after each event. Because they are unknown to Eve before she is able to mount her PNS attack, Alice and Bob can compare frequencies of multi-photon events and very simply detect the presence of an eavesdropper performing a PNS attack. In hardware, this is often implemented using electrically-controlled variable attenuators to modify the coherent attenuation in real time. However, in one of the original demonstrations fixed decoy and signal sources were set and a fiber switch was used to randomly dither between them [Peng et al., 2007].

The beauty of the decoy state protocol is that, by mitigating the PNS attack, it allows Alice and Bob to increase their mean photon number and thus achieve acceptable key generation rates even at long distances or in the presence of significant loss. This technique enabled a demonstration of free-space QKD link over 144 km, well beyond the distance a non-decoy QKD system would be able to transmit securely with mean photon number of 0.1 [Schmitt-Manderbach et al., 2007]. As we will see in later sections, link loss and distance are two key system parameters that limit the performance of real-world QKD systems. Hence any techniques that allow our heros, Alice and Bob, to overcome these limitations are invaluable.

6.2   NONLINEAR AND QUANTUM OPTICAL SOURCES

Weak coherent sources are undoubtedly the most common types of sources in real world QKD links because they are based on the ubiquitous diode laser found in most communications laboratories. However, as systems become more advanced and security and efficiency requirements dominate the design of QKD links, sources based more directly in quantum mechanics will become more important than ever in achieving real world quantum cryptography. Of course, because they are much more difficult to make and tend to become research projects in their own right, quantum optical sources are much less common and much less developed, and we will briefly cover some of the current approaches. Keep in mind that this is one of the most rapidly advancing areas of quantum information research; this treatment will only scratch the surface of the subject and may even be out of date by the time you have reached this chapter of the book.

One common theme among most of the approaches is the employment of nonlinear optical processes, most commonly parametric downconversion or four-wave mixing. These two nonlinear processes result from the nonlinear optical response of certain materials to input light fields. Recall that one can treat light traveling through a medium as being constantly absorbed and reemitted by the constituent atoms of the material, whether a crystal, liquid, gas, or amorphous solid. We commonly consider only the process under which the wave is reemitted at the same frequency as the incoming, absorbed wave. This is the dominant *linear* optical response of most transparent materials. However, because of the asymmetrical properties of the constituent atoms or molecules in most media, often materials will have nonlinear responses as well that will induce them to reemit at various multiples or combinations of the input frequencies. For example, in the process of parametric downconversion, a material such as lithium niobate ($LiNbO_3$) or β-barium borate (BBO) will absorb light at a given angular frequency ω_0 and, via its nonlinear response, reemit at frequencies ω_1 and ω_2 such that $\omega_1 + \omega_2 = \omega_0$. Of course, because these nonlinear responses are usually small, very little of the ω_0 light is actually converted and much of it is still transmitted through the material as well (though this is not always the case). In another process, called four-wave mixing, two input frequencies, ω_1 and

ω_2 are converted to two different output frequencies ω_3 and ω_4 such that $\omega_1 + \omega_2 = \omega_3 + \omega_4$. This higher-order nonlinearity, though often much weaker than parametric down-conversion, is present in all materials, even amorphous glasses. The process of phase matching controls the frequencies where these processes occur

Obviously, more depth into the field of nonlinear optics, while integral to a lot of work in quantum optics, is well beyond the scope of this text. Thus for further information the reader is strongly referred to [Boyd, 2008].

6.2.1   HERALDED SINGLE PHOTON SOURCES

One evolutionary improvement over statistical single photon sources is the development of the heralded photon source. In this setup, single photons from a weak coherent source passes through a nonlinear crystal and is spontaneously converted to two photons of longer wavelength through parametric downconversion. This process, called spontaneous parametric downconversion (SPDC), for obvious reasons, results in two photons for each converted single photon (with some sub-unity efficiency, of course). One way to improve the efficiency of single-photon communications is to use one of these pairs to *herald* the existence of the other photon. This allows Alice and Bob to know which pulses from the weak coherent source actually contain a photon and which ones are empty. With this knowledge, they can avoid having to perform unnecessary post-processing on empty quantum channel pulses.

6.2.2   DETERMINISTIC SINGLE PHOTON SOURCES

While the decoy state protocol outlined above minimizes the impact of the PNS attack, it remains a statistical approach and still requires some sacrifice in both the key reconciliation process and the overall efficiency of the link. Similarly, while heralded photon sources can alleviate some of the processing burden, they still result in 9 out of every 10 pulses being empty. Thus the ideal source for true BB84 or B92 QKD is a deterministic single photon source. In other words, Alice would ideally like to use a source where she provides a single trigger and obtains, with a high degree of certainty, one and only one photon at her transmitter. Of course, this is not such an easy task with today's state of the art. A number of approaches have been pursued and abandoned in recent years. The entirety of a 2004 issue of the New Journal of Physics [Grangier *et al.*, 2004] is dedicated to so-called single photons on demand and included articles about a number of different techniques.

One approach of Migdall, Branning, and Castelletto [Migdall *et al.*, 2002] involves manipulating the statistics of weak coherent sources using nonlinear parametric downconversion. The system, shown in Figure 6.2, consists of arrays of heralded photon sources, where multiple independent weak coherent sources simultaneously generate their own pairs of photons at longer wavelengths through SPDC. Each source sends its heralding photon to a detector and associated central logic that keeps track of which source has generated a pair of photons at a given time (recall that each source still generates mostly empty pulses). The other half of the photon pair is sent to a fiber-optic delay line, with each source associated with a different duration of delay, and finally into an Mx1

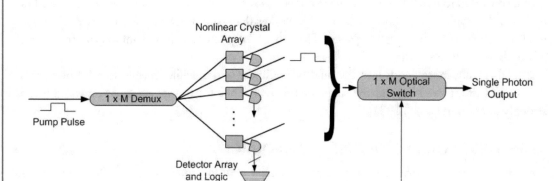

Figure 6.2: *The semi-determistic approach of Migdall, et al., where many individual heralded photon sources are multiplexed to create a single source with more desirable photon statistics.*

optical switch. The logic then controls this switch, routing each delay line to the output in such a way that most of the pulses will contain a single photon from one of the heralded sources.

This approach, while still based in the Poisson statistics of weak coherent sources, achieves a significant improvement in efficiency. Of course, the magnitude of this improvement depends strongly on the number of independent sources used, making the system rather expensive to implement. In addition, Mx1 optical switches tend to have large switching times and can severely limit the overall speed of the source.

6.2.3 BULK OPTICAL ENTANGLEMENT SOURCES

As we learned in Chapter 4, the only truly secure way to do quantum cryptography without the limitations of side channels or random number generators is to use entangled photon pairs. Thus one of the most active areas of quantum communications research is the development of good sources of entangled photon pairs. One of the most ubiquitous and effective approaches, attributable to Paul Kwiat of the University of Illinois at Urbana-Champaign [Kwiat *et al.*, 1999], uses SPDC to create polarization entangled pairs. The apparatus consists of two thin, orthogonally oriented second-order nonlinear crystals. Because of the phase matching requirements, when pumped with diagonally polarized light, this pair of crystals will automatically generate photon pairs that are in the entangled polarization state $HH + e^{i\phi}VV$. The entanglement comes from the fact that, when properly oriented, the polarization of the output photons becomes indeterminate. That is, because the incident light is polarized diagonally to both crystals, the spontaneous downconversion can occur with equal probability in both crystals. Since there is no observable to determine which photon came from which crystal orientation, they become polarization entangled. The arbitrary phase factor $e^{i\phi}$ can be controlled via careful choice of phase matching, crystal thickness, and orientation.

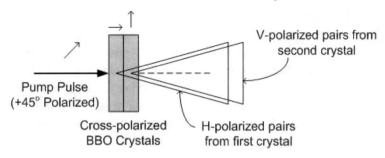

Figure 6.3: *Concept of the bulk entanglement source based on SPDC in a pair of nonlinear optical crystals. The photon pairs emitted at the same solid angle are polarization-entangled.*

6.2.4   FIBER-BASED SOURCES OF ENTANGLEMENT

While bulk-optical systems currently comprise the majority of entanglement generation experiments, they suffer from the same drawbacks as most free-space optical setups, including a strong sensitivity to temperature, vibration, and interference from stray light. Thus bulk sources, while adequate for laboratory or demonstration environments, are in many situations ill-suited when intended for real-world use. Thus there is a strong motivation to pursue sources based in fiber-optics or integrated semiconductor photonics. There are a multitude of approaches under investigation along these lines and all of them certainly cannot receive their due here. However, a few stand out as strong potential candidates for future development into commercial devices. These devices share some commonalities, specifically in the types of nonlinearities that they exploit. Many of the most promising approaches use third-order nonlinear effects such as partially degenerate four-wave mixing, described above, to create entangled photon pairs. Third-order nonlinear processes differ from second-order ones such as SPDC in that they have to potential to create photon pairs with much smaller wavelength shifts. For example, SPDC converts a photon at one wavelength to two photons with longer wavelengths. Because the process must conserve the photon energies throughout, the wavelength shifts involved must be some large fraction of the original wavelength. Thus it is common to have a second-order system that operates through multiple bands in the optical spectrum, converting between combinations of visible, infra-red, and ultraviolet photons.

This potentially large spread in operating wavelengths can pose a severe challenge when using optical fibers or photonic waveguides. Because these devices almost always employ waveguides, their designs are heavily wavelength dependent. Thus to design a waveguide that can operate across broad wavelength bands poses a significant challenge and often leads to less-than-ideal design choices such as significant loss or strong birefringence. For this reason, most waveguide implementations choose to exploit third-order nonlinear effects, where the wavelength shifts can be as small as a few nanometers, to ease design constraints and make for a more effective source.

Of course, with the benefit of small wavelength shifts comes a number of disadvantages. First of all, third-order nonlinearities tend to be by their nature significantly weaker than second-order ones; they require brighter pump sources to generate single photons. Hence, third-order systems are generally less efficient than their second-order cousins. In addition, exploiting one third-order nonlinear effect comes at the cost of having to deal with other third-order effects such as Raman scattering. Raman scattering, which results in the generation of sidebands due to photon-phonon interactions within a material, often generates photon pairs at similar wavelengths as four-wave mixing but that are not entangled, leading to interference and a dilution of the quality of the entanglement. A number of approaches, a sampling of which is outlined below, exist to overcome these limitations and others are surely under development at the moment.

One approach of Alan Migdall and colleagues at NIST involves using the third-order non-linearity in microstructured optical fiber as described above, where a pair of photons at two closely spaced wavelengths are converted via four-wave mixing (FWM) to a pair of photons at a single, degenerate wavelength [Fan *et al.*, 2007]. Through the judicious use of mirrors and beamsplitters, these photons emerge in a polarization entangled state from the system, which is shown in Figure 6.4. The microstructure fiber results in a fundamental transverse mode that is on the order of 1 μm, significantly smaller than the mode size in traditional single-mode optical fibers. This increases the intensity and thus the nonlinear pair generation efficiency of the FWM process. In addition, even though the third-order nonlinear response of ordinary silica glass is rather weak, the low-loss characteristics of optical fibers allows the use of long lengths (on the order of meters to kilometers) in order to increase the overall pair generation efficiency.

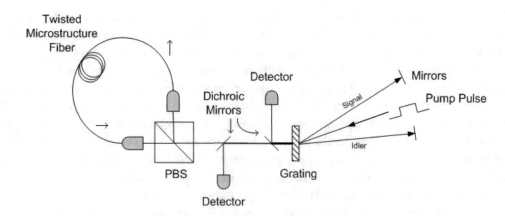

Figure 6.4: *The apparatus of Migdall, et al., to generate polarization entangled photons using FWM in a microstructured fiber. Competition from other third-order nonlinearities makes this approach challenging.*

One challenge in all fiber-based systems is the existence of Raman interference, as described above. In optical fiber, the Raman issue is particularly virulent as the phonon response of amorphous glass is very broadband, creating interference across a large range of wavelengths. A number of techniques have been demonstrated that can attempt to mitigate this problem. One involves cooling the optical fiber by submersing it in liquid nitrogen [Lee *et al.*, 2006]. Since the phonon response of the material is strongly temperature dependent, cooling the fiber decreases the Raman response and eliminates the interference. However, the logistical challenges associated with cooling fiber in a cryogenic liquid tend to outweigh any advantages in practicality gained by using optical fiber in the first place.

Another challenge unique to microstructure fiber is the ability to couple light into and out of the fiber itself. While the compressed mode size increases the nonlinear response of the material, it poses a significant challenge to the coupling optics to focus to a spot of that size with the appropriate numerical aperture. Additionally, the ability to fusion splice that is so advantageous in traditional fibers is lost in these exotically structured fibers, eliminating one of the best coupling methods available in fiber-based systems. Still, fiber-based entanglement generation remains an active area of research, aided by the telecommunication industry's large investment in the underlying technology.

6.2.5 PHOTONIC SOURCES OF ENTANGLEMENT

Much like fiber based systems, entanglement sources based on integrated photonic devices comprise a very promising avenue toward commercially viable quantum communications. These devices typically combine the nonlinear responses of either semiconductors such as gallium arsenide (GaAs) or optical crystals such as lithium niobate with the confinement and manipulation associated with integrated photonic waveguides to achieve good pair generation efficiency in a compact and robust package. While they certainly pose their own challenges associated with devices on such a small scale, they show significant potential as fieldable devices for QKD.

Integrated photonics in the commercial marketplace are most often associated with diode lasers and semiconductor optical amplifiers so ubiquitous in today's telecommunications infrastructure. However, these active semiconductor devices, while useful as pump sources and the like, are not quite the same as the passive sources required for single photon and entanglement generation. This disparity arises from the fact that active devices, where nonlinearities are induced via the interband effects of electron-hole generation in the semiconductor, have associated noise processes that severely mask any underlying quantum effects. Thus all of the semiconductor-based photonic sources of single or entangled photons rely on intrinsic, intraband nonlinearities such as the Kerr effect, which have a much faster and cleaner response than their interband counterparts. Similarly, waveguides fabricated out of nonlinear optical crystals such as PPLN employ the inherently strong, usually second-order nonlinear effects for which they are so well known in their bulk form. Some important examples of such systems are described in [Yoshizawa *et al.*, 2003].

6.2.6 QUANTUM DOTS, SINGLE IONS, AND OTHER NOVELTIES

Quantum dots are currently one of the most rapidly advancing topics in semiconductor research today, and their utility as sources of single photons is one of the main drivers of that research activity.

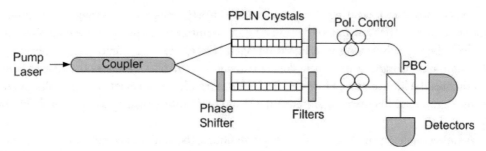

Figure 6.5: *A PPLN-based source of entangled photons. This system exploits the unusually high nonlinear response of lithium niobate along with the advantages of periodically poled phase matching.*

Quantum dots are semiconductor junctions that are confined on such a small scale that they behave in a purely quantum mechanical nature. That is, instead of having an effectively continuous band of energy levels normally associated with semiconductors, material layers in these devices are thin enough to have only a few distinctly separate levels; their behavior mimics that of confined electrons in a single atom or particles in a quantum mechanical harmonic oscillator. For this reason, their transitions can be exploited to generate very well defined single quanta of energy and thus make them a good candidate for single photon sources. However, methods to reliably fabricate quantum dots in a deterministic way still eludes the community. Additionally, quantum dots must be operated at cryogenic temperatures, the logistics of which make them unsuitable for field environments.

Another rather exotic approach has been to exploit the well developed techniques of laser cooling and trapping to isolate individual atoms or ions and implant them in a substrate. Similar to the reasoning behind the utility of quantum dots, their well defined energy levels mean that creating well controlled quanta of energy is a possibility. For example, MacClelland, et al., demonstrated in [McClelland and Hanssen, 2006] the ability to trap and implant a single erbium ion onto a substrate. This effectively created a single-atom laser that would emit only one photon at a time. However, this approach and ones similar to it require large investments in laser cooling experiments that are very sensitive and pose a challenge to recreate outside of the laboratory. While exotic, these single atom sources of photons remain very far from becoming practical sources for quantum communication.

6.3   SINGLE PHOTON DETECTORS

As important as single photons are to the QKD process, equally vital is the ability to reliably detect those single photons on the other end of the link. While the field of single photon detector development is arguably further along than its source-focused companion, there remain a number of challenges. Efficient response across a broad range of wavelengths is one important avenue; a single-photon detector that operates reliably in the telecommunication band without the need for cryogenic cooling remains a 'holy grail.' Broadband QKD also requires detectors with very fast timing characteristics to operate both efficiently and securely. However, unlike source research that is still identifying the most promising path across a broad range of approaches, detector research has proceeded to the point where it is more strongly focused on furthering developments of a relatively small number of basic technologies rather than exploring which of these technologies has the most promise. This section will explore those central detection technologies.

6.3.1 PHOTOMULTIPLIER TUBES

For decades leading up to the development of quantum communications, photomultiplier tubes (PMTs) were the instrument of choice for detecting individual photons. In fact, Bennett's original QKD demonstration was based on PMT technology. These devices are still ubiquitous in many arenas of quantum and nonlinear optics, but they have not become the device of choice in most QKD experiments. The reasons for this are various, but certainly PMTs do not typically have adequate efficiency for QKD at the wavelengths of most common interest, such as 1550 nm or 1310 nm.

On the other hand, PMTs do enjoy a few unique advantages over other single photon detection technologies. One metric in which they surpass many other devices is their timing resolution. As we will delve into in later sections, timing resolution ultimately determines the achievable key rate. If there is a significant spread in the distribution of time differences between when a photon is incident on the detector and when that detector emits an electrical pulse (often called a long detection 'tail'), then the system will suffer from strong inter-symbol interference. This interference can in turn overwhelm the key distillation process and paralyze the secret key generation. Takesuye, et al., encountered this precise problem in developing a high-speed, phase-encoded QKD link [Takesue *et al.*, 2006]. Had they implemented their link using PMTs instead of semiconductor detectors (see the next section) for this particular experiment, they may have been able to distill a secret key using a 10 GHz transmission rate.

6.3.2 AVALANCHE PHOTODIODES

By far the most ubiquitous technology in single-photon experiments is the silicon avalanche photodiode (APD). Usually, these semiconductor detectors are associated with low light applications where their high linear gain is desirable to good signal strength. APDs can be thought of as the semiconductor equivalent of a photomultiplier tube; a typical device structure is shown in Figure 6.6. They consist of an absorption region, a semiconductor junction that absorbs photons at energies larger than the band gap and generates electron-hole pairs as a consequence. The resulting current travels through an amplification region in the semiconductor, where it induces the creation of other electron-hole pairs in what is called an *avalanche*. This results in the high linear gain normally associated with APDs. Furthermore, when biased properly so that a single electron-hole pair from one absorbed photon is enough to cause a cascade of current to form in the amplification region, the device is said to be in *Geiger mode*. Operating similarly to a Geiger counter used to detect nuclear radiation, this device emits an electrical pulse (analogous to the Geiger counter's "click") for each

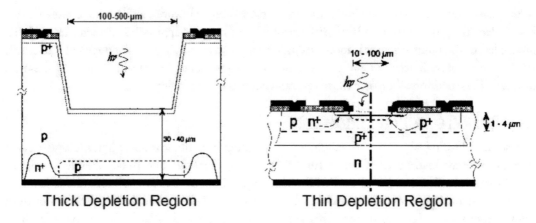

Thick Depletion Region   Thin Depletion Region

Figure 6.6: *A cross-sectional diagram of a typical SPAD (left), showing a thicker depletion region, and the thin SPAD (right) of Cova, et al., where the depletion region is thinned by as much as ten times.* (From S. Cova, M. Ghioni, A. Lotito, I. Rech, F. Zappa, "Evolution and prospects for single-photon avalanche diodes and quenching circuits," *Journal of Modern Optics,* **51** ©2004 reprinted by permission of the publisher (Taylor & Francis Group, http://www.informaworld.com).)

single-photon. Single-photon APDs, or SPADs, are usually combined with back-end circuitry that outputs a TTL-level pulse for each current spike from the diode. The circuitry also controls the *quenching* process whereby the electron-hole pairs are flushed from the amplification region and the device is readied to absorb another photon.

A number of characteristics of SPADs are important to the successful design of QKD systems. Two obvious detector parameters that will affect a link's performance are detection efficiency and timing resolution. Single photon detectors generally operate by generating a TTL or similar logic pulse for every photon that is absorbed. Of course, the detector's absorbing material only absorbs with some sub-unity efficiency, so not every incident photon will result in an output pulse.

Conversely, SPADs can emit what are called *dark counts* or pulses that are not associated with incident photons but rather due to thermal noise resulting from the fact that they are operated at some non-zero temperature. These dark counts look to a QKD receiver like noise and must be filtered out in the sifting process, putting further strain on QKD post-processing. Thus detection efficiency and dark counts are two important parameters in judging whether a SPAD is appropriate for QKD.

Timing resolution is another important feature in choosing SPADs that are appropriate for broadband QKD links. The ability for a detector to resolve the difference in time between two adjacent single photon events ultimately limits the performance of any broadband QKD receiver. Thus, having detectors that are capable of recovering quickly from one photon event in order to be able to detect another are crucial for practical quantum communication. Typically, the SPAD characteristic that most strongly affects its timing resultion is the volume of the device's absorption region. Because an incident photon can be absorbed anywhere within this volume, the time between its absorption and the resulting electrical pulse has some random, statistical distribution with non-zero variance, referred to as *jitter*. Typical SPADs can experience timing jitter in excess of 200 ps. For systems operating at GHz clock rates, this number can be very significant and often limits the overall system clock speed to avoid inter-symbol interference. Recent efforts of Sergio Cova and colleagues at the Politechnico di Milan [Gulinatti *et al.*, 2005] have worked to reduce this timing jitter by fabricating devices with significantly smaller depletion regions, as shown in Figure 6.6. While a smaller depletion region results in a detector efficiency that is both reduced overall and shifted into the visible part of the spectrum as shown in Figure 6.8, the timing jitter of such a device is reduced to 35 ps as shown in Figure 6.7. The tradeoffs and utility of using such detectors in broadband QKD links is discussed in the following chapter.

As mentioned above, the quenching circuitry on a SPAD controls the the process of clearing out the charge carriers from the amplification region and readying the device for another single photon event. While the time that this process takes has obvious ramifications for the maximum count rate of the detector, it also has another important implication: security. Work performed at NIST demonstrated that detectors with finite recovery times (also known as the detector's *dead time*) have strong implications in BB84 and similar protocols [Rogers *et al.*, 2007]. Because the protocols rely on an equal probability of detection across all of the detectors, the possibility that any one of them is disabled for some time greater than the transmission period instantly skews the statistics of the seemingly random key. In BB84 QKD with typical detectors, this effect can result in keys that are correlated in excess of 60%, more than 10% higher than 50% correlation expected for a truly random key. For detectors with some finite dead time longer than the transmission period (a situation encountered in most broadband BB84 QKD implementations), security requirements dictate that one must modify the sifting procedure to ignore detection events that occur within the dead interval of the other detector in a given basis. Of course, throwing out detection events leads to a further loss of efficiency, so detectors with shorter recovery times are beneficial to high-speed QKD. Alternatively, one can use the modified BB84 receiver architecture shown in Figure 6.10. This so-called self-disabling receiver uses a time-division multiplexing scheme to ensure that both detectors in a given basis are disabled at the same time [Rogers *et al.*, 2007]. Another detector characteristic that affects security is afterpulsing. This phenomenon denotes the reemission of photons from the SPAD as the charge carriers are quenched. The security implications are obvious; were an eavesdropper able to view any optical emission from Bob's BB84 receiver, she would potentially be able to discern which detector fired at a given instant. Combined with her assumed access to the classical channel, this would give Eve all of the information she would need to recover the key. Thus understanding and controlling afterpulsing from a detector is one important consideration when implementing truly secure QKD.

6.3.3   INGAAS AND UP-CONVERSION DETECTORS

Most of the SPADs used in QKD are based on silicon CMOS devices. This is because Si devices offer acceptable efficiency in the near infrared with adequately low dark count rates. However, silicon's absorption spectrum falls off at wavelengths longer than 1100 nm, making them unsuitable for QKD systems that operate at typical telecommunications wavelengths. This is of particular interest for fiber-optic systems, as fiber tends to have higher loss at wavelengths shorter than 1300 nm. Thus there is strong interest in single-photon APDs that operate at wavelengths in the range of 1.3 to 1.6 μm. One approach is simply to use APDs made of other semiconductor materials that are responsive in this wavelength band (often called the telecommunications window). III-V semiconductors have traditionally been suitable for photonic devices operating at telecommunications wavelengths, and a number of III-V SPADs, often made from materials such as InGaAs, have been demonstrated for

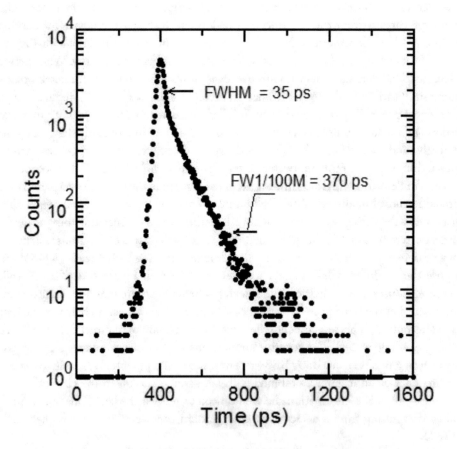

Figure 6.7: *The timing jitter of a thin (1-4 μm) SPAD device, showing a FWHM of 35 ps, compared to over 200 ps in a normal device.*

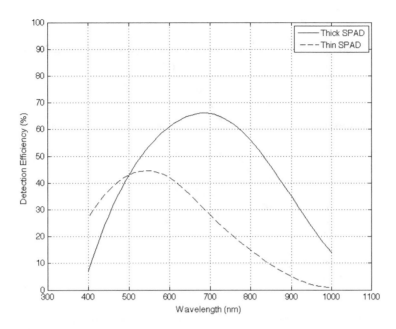

Figure 6.8: *The relative quantum efficiencies of a normal, thicker SPAD and a thin device. Note the overall loss in detection efficiency, as well as the shift of the peak from the near IR into the visible region of the spectrum.*

QKD applications [VanDevender and Kwiat, 2007]. However, one consequence of their increased sensitivity to longer wave infrared radiation is their severely increased dark count rates. Dark counts associated with InGaAs SPADs are often so strong that either they overwhelm the key reconciliation process or they require cryogenic cooling to reduce the thermal radiation in the detector. The requirement of a cryogenic system has made them unsuitable for real-world QKD applications and there is much research into InGaAs SPADs that do not require such stringent temperature control measures.

Another approach to detecting single photons in the telecommunications window is to use yet another nonlinear optical process similar to the ones we learned about in the previous sections. However, instead of converting a shorter wavelength photon to two longer wavelength ones in the process of SPDC, this approach, called *up-conversion*, uses a nonlinear crystal to convert a telecommunication band photon into two photons, one with a shorter wavelength and one with a longer one (their energies must still add up to the original in order to satisfy conservation of energy, of course). In this way, one can use a Si SPAD to detect the shorter wavelength photon that results from the nonlinear interaction of the original telecommunications band photon with the crystal. Of course, there is some sub-unity efficiency associated with the up-conversion process that limits

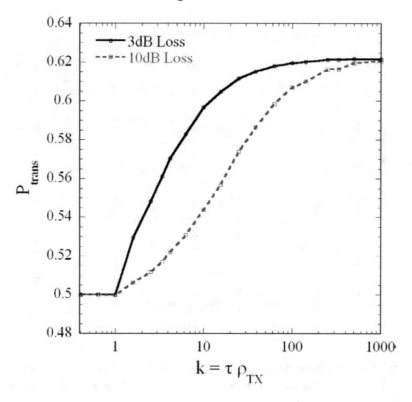

Figure 6.9: *Correlations in a supposedly random key in the presence of non-zero detector dead time. Plotted is the transition probability, P_{trans}, defined as the probability that a bit will be of opposite value than the one preceding it, versus the normalized dead time, defined as the product of the dead time, τ, multiplied by the transmission rate, ρ_{TX}. Note how rapidly it diverges from the expected value of 50%.*

the overall efficiency of this detection system at longer wavelengths. Still, various demonstrations have been able to achieve up to 99% detection efficiency at 1550 nm using up-conversion detectors [VanDevender and Kwiat, 2007] and have shown them to be of great value to fiber-optic QKD systems.

6.3.4   SUPERCONDUCTING SINGLE-PHOTON DETECTORS

One of the most exotic but effective new methods to detect single photons is the superconducting single-photon detector (SSPD) developed independently by Sae Woo Nam at NIST [Takesue *et al.*, 2007] and by Dzardanov in Moscow [Gol'tsman *et al.*, 2001]. This device, depicted in Figure 6.11, consists of a nanoscale meander wire made of niobium nitride deposited on a glass substrate and connected to an external resistance monitoring circuit. The device is cooled cryogenically just below

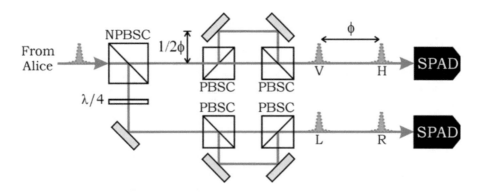

Figure 6.10: *A self-disabling BB84 receiver that is immune to effects caused by the detector dead time. The bits in each basis are time-multiplexed with period ϕ, so each detector is responsible for detecting both bits in a given basis. Thus when one detector goes dead, it disables the entire basis automatically and eliminates any key correlations.*

the critical temperature of NbN so that it becomes superconducting. In fact, the temperature is so close to the critical temperature that a single photon absorbed into the NbN wire is enough to heat it above T_c, causing the resistance of the wire to increase dramatically. In this way, they are able to detect single photons with relatively good efficiency (20%). Also, because their thermal mass is so low, these detectors have extremely short recovery times (10 ps). Of course, there are some unique tradeoffs in these types of devices between efficiency (related to the density of the meander wire path on the substrate) and the timing jitter (related to the amount of time the current takes to traverse the wire from the point of the photon absorption). Once again, the cryogenic requirement is a significant drawback to these types of devices that must be weighed against their efficiency and recovery time.

6.4   CONCLUSION

As we will discover in the next chapter, a QKD system is only as capable as the parts used to construct it. We have surveyed here a number of the most common devices used to implement single-photon communications links. As new devices are invented and existing ones improve, the capabilities of secure single-photon communications will only improve with them.

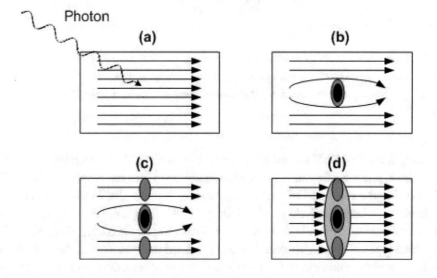

Figure 6.11: *The process of a superconducting nanowire detecting a single photon. (a) The supercurrent propagates along the cooled wire. A photon of sufficient energy is absorbed. (b) The photon energy locally heats a portion of the wire, causing the Cooper pairs to divert around the hot spot. (c) The increased current density further heats the wire, causing (d), a macroscopic change in resistance that is detectible via a resistance bridge circuit.*

CHAPTER 7

A Survey of QKD Implementations

At this point, we have learned about each specific aspect of quantum cryptography, from protocols to key reconciliation to specific devices used to handle individual quanta of light. We are now fully prepared to explore some examples of links that have drawn on the principles of the previous chapters to achieve quantum-encrypted communications. While this chapter is by no means intended to cover all of the links demonstrated (there are far too many, with more published all the time), it is meant to illustrate the various tradeoffs that must be considered when designing a QKD link. Each link presented in this chapter was designed with a different focus in mind, whether it be high speed, long distance, or enhanced security. We will learn how those various foci impact the design choices of each link.

Very broadly, one can divide the field of experimental QKD into two categories: free-space and fiber-optic form factors. Each one has a different set of associated requirements and challenges. For example, a photon's polarization is very stable under free-space propagation but very unstable in optical fiber. Similarly, free-space systems must contend with interference from background light and loss due to atmospheric scintillation, while fiber-optic systems are obviously immune to such effects. Thus, each form factor is suitable for a different application, as we will discuss in the this and the next chapter. In the meantime, let us explore a few examples of each.

7.1 FREE SPACE QKD SYSTEMS

The diverse implementations of free-space quantum key distribution all share some commonalities. For example, they all must contend with interference due to background light. Often the easiest way to cope with background light (primarily originating at the Sun[1]) is to operate the link only at night. Of course, while perfectly suitable for a proof-of-concept demonstration, this requirement is not practical for any real-world system. Other ways to eliminate solar background noise include strong spatial, spectral, and temporal filtering or even more exotic approaches, like choosing an operating wavelength that falls into one of the dark Fraunhofer lines in the solar spectrum [Rogers *et al.*, 2006].

Free-space QKD systems must also contend with all of the challenges that arise in classical free-space optical communications as well. Propagation loss and alignment issues are two dominant challenges that impede link performance. However, there is one major difference between classical

---

[1]A simple calculation reveals that the Sun bombards each square meter of the earth with over 10^{21} photons each second. Finding one photon among all of those is no easy task.

and quantum optical links: In classical links, one can often compensate for increased loss or misalign-ment errors by simply transmitting more optical power. However, for obvious reasons, free-space QKD transmitters must be limited to the single-photon level. In order to achieve the desired link performance, one can only compensate by sending photons more frequently or else becoming more clever in handling the photons that are already there.

7.1.1   BROADBAND LINKS

Of the many performance metrics driving the design of QKD links, key generation rate lies among the most important. Any practical QKD link suitable for integration into the modern infrastructure must operate at speeds commensurate with broadband communications. At the very lowest limit, this means that a truly broadband quantum cryptosystem must generate key bits at greater than 10 Mb/s to be compatible with first generation ethernet protocols. Currently, this remains an open challenge to the community (though that statement may no longer be true by the time you read it!). In the interim, only a few links have achieved key rates that are close to this goal. To achieve such high key rates, these links generally rely on techniques borrowed from classical telecommunications engineering. For example, the link from the group at the National Institute of Standards and Technology (NIST) in Gaithersburg, Maryland, achieves secret key rates in excess of 1 Mb/s using a clock recovery scheme on the classical channel, allowing for careful synchronization of the transmitter and receiver. By employing conventional 8-bit/10-bit encoding of the classical channel data, they ensure that there are enough bit transitions on the classical channel to reliably recover Alice's transmitter clock. In turn, this allows Bob to sharply limit the window in which he allows his detectors to fire, significantly reducing background noise counts. In this system Alice employs high-bandwidth vertical cavity surface emitting lasers (VCSELs) as her weak coherent source. Because of their unique structure, VCSELs have a low threshold current and subsequently very high modulation bandwidth. By rapidly pulsing small amounts of current through the VCSELs, Alice is able to clock her single photon transmissions in excess of 1 GHz. Finally, the sifting, error correction, and privacy amplification algorithms are optimized through judicious use of multithreaded computer code and custom

Of course, this primary focus on key rate does have some disadvantages. It is a BB84 system and therefore relies heavily on a good random number generator. In the case of the NIST design, the PRNG is its weakest element; to achieve such high speeds, it demands random bits at a very high bit rate. The NIST system in [Bienfang *et al.*, 2004] uses a Mersenne twister that, as we learned in Chapter 2, is not cryptographically secure. In order to improve the security, the system would have to employ some type of fast quantum random number generator. Alternatively, employing these similar telecommunications engineering techniques to an entangled QKD system may lead to a truly broadband system with the ultimate, entanglement-based security.

7.1.2   LONG DISTANCE COMMUNICATIONS

While the NIST system focuses on speed, another link sets the standard in terms of transmission distance. The link of Anton Zeilinger and colleagues [Schmitt-Manderbach *et al.*, 2007] achieves

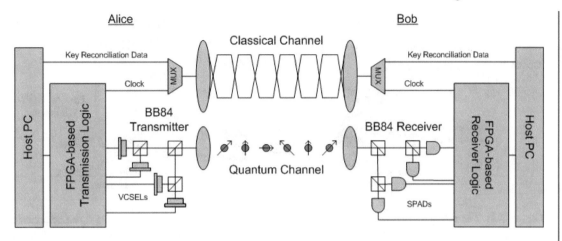

Figure 7.1: *The broadband QKD link of NIST. This system uses clock recovery on the classical channel to synchronize Alice's transmitter to Bob's receiver to much greater resolution than the clock period, allowing for very sharp temporal filtering. Additionally, Alice uses VCSELs as her weak coherent source, enabling her to modulate her transmitter at GHz rates.*

secure key distribution over a distance of 144 km. Set up between the two Canary Islands of Tenerife and La Palma, the link uses as its receiver telescope an optical ground station owned by the European Space Agency. Across this link the group has implemented both decoy-state and entanglement-based QKD, successfully demonstrating the longest-range QKD link in existence to date. The distance was chosen to mimic the types of ranges required to communicate with a low-earth orbit (LEO) satellite. However, in the demonstration the group only achieved a secret key rate of 13 bits per second, hardly close to a rate required for a practical communication scenario. Still, this high-visibility experiment has gone a long way in demonstrating a path to global, satellite-based QKD, as we will discuss in the final chapter.

7.1.3   TRULY ENTANGLED LINKS

Most of the free-space QKD demonstrations discussed so far have implemented some form of the BB84 or B92 protocols, and we have already learned that only entanglement-based QKD achieves the promise of ultimate security. However, because entangled photon pairs are difficult to create and exploit, not many E91 QKD links have been demonstrated. One exception is the link of Christian Kurtseifer, et al., at the National University of Singapore (NUS) [Marcikic *et al.*, 2006]. The NUS link uses a continuous source of entangled photons created by SPDC in a BBO crystal, as shown in Figure 7.2. Using this source, Alice transmits one half of the pair across a 1.5 km free space link propagating over the city. She then keeps the other half of the pair and sends it to a time-stamped single-photon detector. Because the source is continuous, it generates single-photon pairs

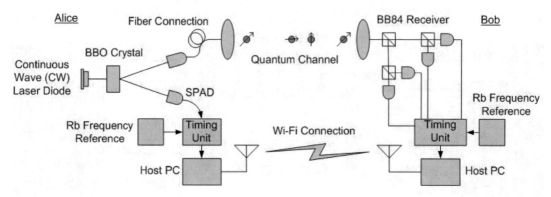

Figure 7.2: *The entangled photon link from the National University of Singapore. This system uses a continuous source of entangled photon pairs that is post-selected based on synchronized atomic clock time stamps to achieve a key rate of 1 kb/s.*

stochastically, with exponentially-distributed times elapsing between photon pair events. Her time-stamping detector then records the time at which each half of a pair was detected referenced to a rubidium oscillator. At the other end of the link, Bob receives the first half of the photon pair and measures its polarization with his own time-stamped single photon detector slaved to a second rubidium oscillator and synchronized with Alice's. Because rubidium atomic clocks are so accurate, Alice and Bob are able to correlate the detection times of their respective detection events and match up which events belonged to which pairs. By post-processing these pairs in the usual way, they can distill a secret key via the E91 protocol. The accuracy of the atomic clocks at both ends are what allow Alice and Bob to use a continuous, stochastic source of entangled photons, rather than the separately synchronized pulsed source that is often associated with other E91 QKD systems.

Even with this advanced entanglement-based system, the NUS group is only able to achieve a generation rate on the order of 1 kb of secret key per second. Though they are able to achieve truly entangled security, they have a significant way to go before being able to use their QKD link for broadband data transmission applications.

7.2   FIBER OPTIC QKD SYSTEMS

The other class of QKD systems relies on dedicated optical fiber to transmit the quantum channel information rather than the free-space optical path used in the links outlined above. Fiber-optic QKD poses an entirely different set of challenges than free-space QKD. Most significantly, the polarization encoding used in free-space systems is highly unstable in optical fibers. Any slight change in the stress the fiber can cause severe changes in the photon polarization propagating though it. Obviously, this is a significant problem in applications where the dedicated fiber would lay under streets or hang from poles; vibrations and environmental effects would introduce highly variable and unpredictable strain. Much of the work in fiber-based QKD goes into either stabilizing the photon polarization in the fiber or in coming up with an alternative encoding scheme that exploits a different quantum mechanical property of single photons altogether.

Additionally, glass optical fibers are only low-loss (typically less than 0.5 dB/km) in certain windows. As mentioned in the previous chapter, these windows lie roughly between 1.3 and 1.6 μm where it is difficult to make good single photon detectors. Outside of this range of wavelengths, standard optical fiber can have losses that are 4-5 times higher than those encountered in the telecommunication bands. Thus choosing a wavelength that is compatible with Si detectors means limiting the distance over which the single photon can propagate, while choosing a good wavelength for the photon to propagate in the fiber requires working with less efficient or less practical detectors.

7.2.1   COMPENSATED POLARIZATION ENCODING

The most straightforward way to implement fiber-optic QKD is simply to replace the free-space optical path between Alice and Bob with a fiber one. However, this will only work over short distances with fibers that are clamped in place and do not experience any variation in stress or temperature. Thus a bit more effort is required to make such a link operate reliably. One approach, also from the QKD group at NIST, implements an active polarization compensation system that periodically measures the change in polarization state due to the optical fiber throughout the key distribution process and adjusts electrically-controlled rotators spliced into the path in order to compensate for variations [Ma *et al.*, 2006]. This method, though less elegant, performs rather well. However, this link still uses single photons around 810 nm, subjecting the link to higher propagation loss and thus limiting it to relatively short distances.

7.2.2   PHASE ENCODING

As mentioned above, polarization is a poor variable to use for quantum coding when propagating single photons through an optical fiber. Thus many approaches seek to code qubits on other states of the single photon. One of the most common alternatives is phase encoding, where random bits are coded on the relative phase states of the single photons. There are a number of reasons that this is more suitable to a fiber-optic transmission medium. First, electrically-controlled phase modulators are readily available telecommunications components, so manipulating the phase of single photons is a well-established technique. Second, a photon's phase is much more stable than its polarization when propagating through a fiber. By applying simple phase-shift keying techniques borrowed from classical telecommunications, one can encode qubits on photons in a way that is much more suited to a fiber-optic transmission path.[2]

There is one major drawback to most phase-coded QKD systems. Because phase must always be measured relative to some reference signal, Alice must also transmit a reference beam to Bob in order to allow him to make meaningful phase determinations. Hence this scheme requires not

---

[2]One begins to notice the pattern of quantum communcations borrowing heavily from classical communications, only applied to much lower signal levels.

only dedicated quantum and classical channels, but also a phase-stable reference signal as well. Additionally, even when Alice is able to transmit a reference signal, phase perturbations from the system or the environment can make it difficult to phase-lock both ends of the link reliably.

One of the most straightforward implementations of this encoding scheme was demonstrated by Townsend, Rarity, and Tapster in [Townsend *et al.*, 1993]. Their setup, shown in Figure 7.3, essentially extends a Mach-Zender interferometer over as long as 10 km of optical fiber. This eases the requirement of stabilizing the reference arm since most of the interferometer shares a common propagation path and thus any perturbations are canceled out. The only coherence requirement is that the source be phase-stable for the time-equivalent of the path difference, which is only a few meters on each end of the link.

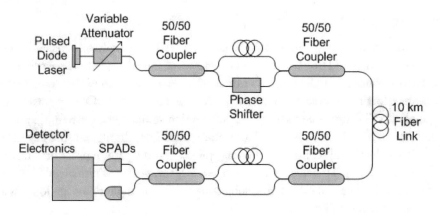

Figure 7.3: *A simple phase-coded QKD implementation, where the reference signal and phase-shifted photons are transmitted on the same fiber.*

The group of Yoshi Yamamoto at Stanford University has worked around the phase stability issue even further by applying differential phase keying (DPSK) rather than simple phase modulation and reference comparison of previous approaches [Diamanti *et al.*, 2006]. In their system, Alice splits each optical pulse from her source into three time-divided sub-pulses, modulating her random bits only onto the phase of the middle pulse, as shown in Figure 7.4. Taking advantage of the fact that most coherent optical systems are phase-stable at intervals shorter than the coherence time, Bob then uses the first and last sub-pulses to determine the relative phase of the center pulse. This technique, again borrowed from classical telecommunications engineering, eliminates the need for a reference signal and makes fiber-optic phase-coded QKD a much more practical system.

7.2.3   SINGLE-SIDEBAND FREQUENCY CODING

Of all of the fiber-based approaches to QKD currently under development in the community, frequency encoding stands out as the most promising for real-world application. Demonstrated by

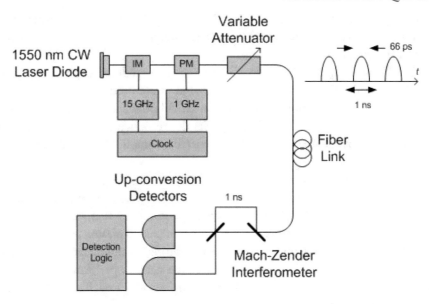

Figure 7.4: *The DPS-QKD setup of Yamamoto, et al. Each bit is split into three pulses, where the center one contains the phase-coded data. The framing pulses provide the phase reference. Since everything shares a common path, the interferometer does not need to be stabilized and the source only requires nanosecond scale phase coherence.*

the group of Jean Marc Merolla [Bloch *et al.*, 2007], this coding scheme takes advantage of yet another, even more stable quantum aspect of single photons: their optical frequency.

In this design, single photons are generated by strongly attenuating a modulated coherent source. This results in single photons whose frequency spectrum consists of a center peak (the photon's frequency) as well as modulation sidebands. By controlling the phase of the modulation, Alice is able to prepare the photon with a single modulation sideband either above or below the photon's center frequency. These sidebands then act in the same way that any two-state quantum system would behave. When Bob measures the phase of the modulation using his own optical modulator, he can choose from one of four phases in two non-orthogonal bases. The process is similar to the quadrature phase shift keying (QPSK) scheme described before, except instead of varying the photon's phase, now one is varying the phase of the *modulation*. This is significantly easier to implement, since synchronizing the phase of an RF signal over long distances is much easier than stabilizing an optical phase reference. In fact, often Alice's and Bob's respective modulators can be synchronized via the existing classical channel. Because the modulation sidebands are shifted in frequency by often only hundreds of MHz, they are not susceptible to chromatic dispersion in the fiber. Additionally, they are significantly more stable when propagating through fiber than either phase, which is susceptible to small path length variations, or polarization, which is sensitive to small changes in stress. For this reason, single-sideband (SSB) frequency-coded QKD is one of the most promising approaches for fiber-based systems.

There is still one unique challenge to SSB coding. The modulation process results not only in sidebands shifted at the RF modulation frequency, but it also produces smaller sidebands at the various harmonics of the modulation frequency as well. While much lower than the fundamental and first lobes, these higher-order modulation peaks do introduce the potential for information leakage and present an exploitable signal if not restricted in amplitude via adjusting the intensity and modulation depth.

One interesting aspect to consider is the possibility of extending this coding scheme for use with an entanglement-based protocol. We have examined various sources of polarization entangled photon pairs but have not considered ones that create photon pairs that are frequency entangled. Often one of the most challenging aspects of creating an entangled photon source from a nonlinear crystal is to convert the conjugate frequency pairs created during the nonlinear process into conjugate polaritons. By using frequency coding, one eliminates the need to perform this extra step in the entanglement source. In this way, frequency-coded, entanglement-based, fiber-optic QKD has a lot of potential as a practical system.

7.3   COMMERCIALLY AVAILABLE SYSTEMS

As with any emerging technology, there are a number of start-up companies that have released commercialized versions of QKD systems. These companies have various target markets in mind, including banking, health care, and government systems. One system was even demonstrated as a way to secure electronic voting in a national election in Switzerland [Swiss Quantum, 2009]. However, all of these systems suffer from one common drawback: They do not implement quantum cryptography in its entirety. Because these systems (or indeed, any system currently in existence) are not fast enough to integrate fully into the modern telecommunications infrastructure, they instead use their quantum key generation systems to swap keys for some other symmetric cipher such as AES (recall chapter 2). Because they do not implement the one time pad, they lose a major component of QKD's claim to ultimate security. While there certainly may be marketable opportunities for QKD (something we explore in the next chapter), many commercial systems available today are rather premature and have the unfortunate effect of confusing or skewing the market's perception of the technology. However, one significant positive contribution from industrial investment is the development of novel or improved devices. While originally intended for quantum cryptography systems, devices such as single-photon APDs are finding their way into spectroscopy and biophysics experiments, opening whole new avenues of research is these relatively disparate fields.

7.4   NEXT GENERATION QKD

In spite of the nascent state of much of the technology, QKD will surely progress as a field well into the future. Along those lines, the links outlined here are presented with the caveat that they are only the most recent examples and may even be obsolete by the time this manuscript goes to press. Hopefully, their presentation can provide a background on which a new generation of researchers can make inspired improvements. In this way, the inevitable obsolescence of the material provided here can only be considered to be the success of that new generation.

CHAPTER 8

Conclusion - QKD in the Marketplace

No discussion of quantum cryptography would be complete without considering how the technology might fit into the overall telecommunications marketplace. We have learned a great deal of technical detail about QKD, but we have yet to examine whether all of this work is worth doing and how it might actually apply in the future. It is often easy to focus only on how incredible it is that fundamental quantum mechanics can be applied in such a dramatic way to such an important problem; in the physics community, we often supplant real-world utility with technical novelty. Just because it is a neat idea does not always mean that it will make a big impact. I would like to close this text with a brief, somewhat less technical discussion of the prospects for quantum cryptography in the future marketplace.

8.1 PERCEPTIONS AND MISCONCEPTIONS

Before we hypothesize on the utility of QKD in the real world, it is imperative that we clear up a few misconceptions that often plague the discussion of quantum cryptography among groups less familiar with the technology. As with any headline-grabbing research, mainstream reports of the fundamentals of QKD are fraught with misinformation or oversimplification that often lead to a lack of understanding. In response to these misconceptions, I outline a few facts about QKD that can hopefully dispel some of the false information in circulation about the technology.

- QKD claims to be unconditionally secure only in the idea that one cannot eavesdrop on the transmitted bits without introducing errors into the transmitted key. The errors introduced are statistical, meaning that there is a chance that individual photons can be intercepted and measured. However, the key reconciliation process can eliminate these compromised bits to statistically insignificant levels. Thus the security is in a statistical, information theoretic sense.

- QKD only claims to secure the space between Alice's transmit aperture and Bob's receive aperture. If an eavesdropper has access to Alice's transmitter or Bob's receiver, then there is nothing that single photon quantum mechanics can do to keep her from stealing the key. Similarly, QKD links must be authenticated by some other means, and they are just as vulnerable to a denial-of-service attack as any other communications link. However, QKD links, when implemented properly, can ensure that a key will not be stolen en route to its intended recipient without the knowledge of its intended owners.

- QKD is only as good as its implementation. That is, if a BB84 system has a poor random number generator or some side channel open to attack, then obviously the system will be vulnerable to attack. *However,* truly entangled QKD systems are unconditionally secure. That is, assuming quantum mechanics is correct (or very nearly correct, of which there is very little doubt), then the mere fact that the system is entangled and verified by the Bell test guarantees that no information can leak from the secret key. In this way, we can claim that QKD does achieve the ultimate physical security of information.

It is hoped that this summary can provide at least some reasonable arguments to often polarizing accounts of quantum information security.

8.2   GLOBAL QKD SATELLITE NETWORKS

Now that we know how to build broadband QKD links, the question remains how they may be useful in a real-world communications scenario. The fundamental requirement driving any practical application of QKD technology is the fact that the quantum channel requires a dedicated optical path. This makes QKD links fundamentally different from typical optical communications links that employ amplifiers, splitters, multiplexers, etc., in order to transmit signals over long distances and distribute them to multiple users. Any quantum-encrypted network, on the other hand, will not only require dedicated fiber or a line of sight communications path but also be limited to simple point-to-point network architectures. From a telecommunications engineering point of view, this topology is both expensive and inefficient. However, a few plausible ideas exist for large-scale QKD networks that may have some promise should the demand arise.

One of those ideas is the concept of a global QKD satellite network, described in [Aspelmeyer *et al.*, 2003, Rarity *et al.*, 2002, Villoresi *et al.*, 2008]. In this concept, depicted in Figure 8.1, QKD links between various ground stations around the globe and a constellation of low earth orbit (LEO) satellites would provide a key distribution network, enabling the one time pad to be used between the ground stations along open, classical telecommunications channels. While expensive and limited only to these ground station hubs, this model is certainly the most appropriate for a global-scale quantum encrypted network. Fiber-optics certainly cannot perform the same task, as without the ability to use amplifiers to boost signal levels, communications would be limited only to small areas and would require many routing stations, threatening security and increasing the cost.

The group of Anton Zeilinger and colleagues has been leveraging their success with long-distance QKD to perform a demonstration of entangled photon distribution using a source on the International Space Station. This project is ongoing and has the potential to boost support for such a network. However, in the end the investment required to implement the network on a global scale would require further advances in the technology, not to mention significantly more support from the end user community, whoever that may be.

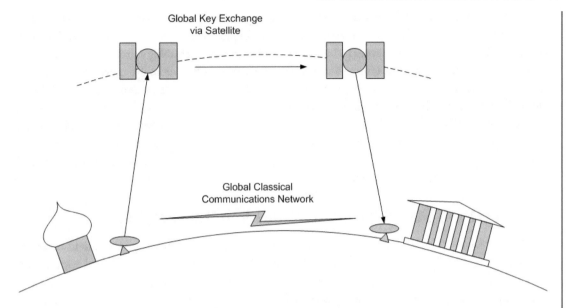

Figure 8.1: *A cartoon of a global satellite-based QKD network. Optical links to secure or trusted satellites would distribute keys around the globe. These keys would then be used to secure one-time-pad encrypted communications over traditional links between remote places.*

8.3   SHORT-HAUL FIBER NETWORKS

While fiber-optic systems may not be suitable for global scale networks, there are certain applications for which they may fit rather well. For example, a highly secure network among bank buildings in a metropolitan area or a secure connection between a hospital and a data center in order to protect electronic health records may demand the level of security that QKD can provide. In these metropolitan area network (MAN) scenarios, dedicated fibers are more straightforward to install and may even exist already from previous infrastructure installations. Whether this constitutes a real market remains to be seen.

Currently, there exist a number of demonstration networks in place, including throughout the city of Vienna [Peev *et al.*, 2009] and around the Boston metropolitan area [Elliott, 2002]. These research networks are useful for exploring how quantum-encrypted MANs may be deployed in future financial districts, among sensitive government buildings, or between various health care facilities to enable the secure transfer of some of the world's most sensitive data.

8.4   THE REAL NEED FOR QKD

All of these speculations, however, beg one overarching question: Do we really need quantum cryptography? All of these concepts about implementing quantum networks on metropolitan or global scales are moot unless a real demand for the technology exists. We know that, in theory, RSA and related cryptosystems are not provably secure. However, in practice, they remain to this day the hardest part of a communications system to break. Most information technology professionals will posit that it is not worth the cost or effort to attack the encryption when there are many other exploits, from social engineering attacks to spyware, that one can employ to breach secure data. Until there are major breakthroughs in either prime factorization algorithms or practical quantum computing, investment in QKD will remain more of an insurance policy against worst-case scenarios rather than a high priority in the information security community. Still, one never knows how technology development will progress, and the knowledge gained in developing QKD will certainly be useful in other areas of science in the meantime.

8.5 THE FUTURE OF QUANTUM ENGINEERING

A colleague once remarked to me that much of the current work remaining in the field of QKD is engineering, but that it is not the kind of engineering that those with a traditional engineering background are well prepared to perform [Jaduszliwer, 2008]. Rather than minimizing the role and capability that engineering plays in modern technology development (for this colleague himself worked more in engineering than in pure science, as do I), I took his statement more to imply that modern engineering training does not emphasize quantum mechanics to a sufficient degree to impart an understanding of the deeper concepts of quantum information. Thus I tend to agree with his assessment in that performing the necessary engineering of quantum cryptosystems will require more in-depth education of the fundamental principles of quantum entanglement and information, topics that are generally confined to upper level graduate physics study. As QKD leads quantum information science into the mainstream of the modern communications infrastructure, I foresee the growth of a new discipline of quantum engineering emerging to provide the skills that this technology demands. As we can see from these previous chapters, the training will combine deep study of quantum mechanics along with classical information theory, cryptography, and electrical engineering. Much like the biotechnology industry that has demanded concomitant understanding of biology, chemistry, statistical physics and information science, this new field promises to bridge a number of disparate subjects and become a truly interdisciplinary area of scholarship.

Bibliography

Aspect, A., P. Grangier, and G. Roger, 1981, Phys. Rev. Lett. **47**(7), 460. 27, 28

Aspelmeyer, M., T. Jennewein, M. Pfennigbauer, W. Leeb, and A. Zeilinger, 2003, Selected Topics in Quantum Electronics, IEEE Journal of **9**(6), 1541, ISSN 1077-260X. 74

Bell, J. S., 1966, Rev. Mod. Phys. **38**(3), 447. 26

Bennett, C. H., 1992, Phys. Rev. Lett. **68**(21), 3121. 34

Bennett, C. H., F. Bessette, G. Brassard, L. Salvail, and J. Smolin, 1992, Journal of Cryptology **5**(1), 3, URL http://dx.doi.org/10.1007/BF00191318. 35

Bennett, C. H., and G. Brassard, 1984, in *IEEE International Conference on Computers, Systems, and Signal Processing, Bangalore, India* (IEEE), p. 175. 30

Bennett, C. H., G. Brassard, and J.-M. Robert, 1988, SIAM Journal on Computing **17**(2), 210, URL http://link.aip.org/link/?SMJ/17/210/1. 39

Bienfang, J., A. Gross, A. Mink, B. Hershman, A. Nakassis, X. Tang, R. Lu, D. Su, C. Clark, C. Williams, E. Hagley, and J. Wen, 2004, Opt. Express **12**(9), 2011, URL http://www.opticsexpress.org/abstract.cfm?URI=oe-12-9-2011. 64

Blake, I. F., G. Seroussi, and N. P. Smart, 1999, *Elliptic curves in cryptography* (Cambridge University Press, New York, NY, USA), ISBN 0-521-65374-6. 9

Bloch, M., S. W. McLaughlin, J.-M. Merolla, and F. Patois, 2007, Opt. Lett. **32**(3), 301, URL http://ol.osa.org/abstract.cfm?URI=ol-32-3-301. 69

Boyar, J., 1989, J. ACM **36**(1), 129, ISSN 0004-5411. 10

Boyd, R. W., 2008, *Nonlinear Optics* (Academic Press), 3rd edition. 49

Brandt, H. E., 2005, Phys. Rev. A **71**(4), 042312. 38

Brassard, G., and L. Salvail, 1994, Advances in Cryptology —EUROCRYPT '93 , 410. 42, 43

Carter, J. L., and M. N. Wegman, 1979, Journal of Computer and System Sciences **18**(2), 143 , ISSN 0022-0000, URL http://www.sciencedirect.com/science/article/B6WJ0-4B55K9J-D/2/036439eff8b0d54d7974c2d5d6997669. 39, 41

Churchhouse, R., 2002, *Codes and Ciphers: Julius Caesar, the Enigma, and the Internet* (Cambridge University Press). 1

Claude Cohen-Tannoudji, B. D., and F. Laloë, 1973, *Mécanique quantique* (Collection Enseignement des Sciences). 26, 45

Clauser, J. F., M. A. Horne, A. Shimony, and R. A. Holt, 1969, Phys. Rev. Lett. **23**(15), 880. 28

Diamanti, E., H. Takesue, C. Langrock, M. M. Fejer, and Y. Yamamoto, 2006, Opt. Express **14**(26), 13073, URL http://www.opticsexpress.org/abstract.cfm?URI=oe-14-26-13073. 68

Diffie, W., and M. Hellman, 1976, Information Theory, IEEE Transactions on **22**(6), 644, ISSN 0018-9448. 8

Einstein, A., B. Podolsky, and N. Rosen, 1935, Phys. Rev. **47**(10), 777. 26

Ekert, A. K., 1991, Phys. Rev. Lett. **67**(6), 661. 35

Electronic Frontier Foundation, 1999, U.S. Government's Encryption Standard Broken in Less Than a Day, Press Release. 12

Elliott, C., 2002, New Journal of Physics **4**, 46, URL http://stacks.iop.org/1367-2630/4/46. 75

Euclid, J. L. H., Sir Thomas Little Heath, 1908, *The thirteen books of Euclid's Elements*, volume Vol II (Cambridge University Press). 5, 6

Fan, J., M. D. Eisaman, and A. Migdall, 2007, Physical Review A (Atomic, Molecular, and Optical Physics) **76**(4), 043836 (pages 4), URL http://link.aps.org/abstract/PRA/v76/e043836. 52

Fluhrer, S., I. Mantin, and A. Shamir, 2001, Selected Areas in Cryptography , 1. 2, 11

Gerlach, W., and O. Stern, 1922, Zeitschrift für Physik A Hadrons and Nuclei **9**(1), 353, URL http://dx.doi.org/10.1007/BF01326984. 26

Gol'tsman, G. N., O. Okunev, G. Chulkova, A. Lipatov, A. Semenov, K. Smirnov, B. Voronov, A. Dzardanov, C. Williams, and R. Sobolewski, 2001, Applied Physics Letters **79**(6), 705, URL http://link.aip.org/link/?APL/79/705/1. 60

Grangier, P., B. Sanders, and J. Vuckovic, 2004, New Journal of Physics **6**, URL http://stacks.iop.org/1367-2630/6/i=1/a=E04. 49

Griffiths, D., 2005, *Introduction to Quantum Mechanics* (Prentice Hall), second edition. 23, 26

Grover, L. K., 1996, in *STOC '96: Proceedings of the twenty-eighth annual ACM symposium on Theory of computing* (ACM, New York, NY, USA), pp. 212–219, ISBN 0-89791-785-5. 12

Gulinatti, A., P. Maccagnani, I. Rech, M. Ghioni, and S. Cova, 2005, Electronics Letters **41**(5), 272, ISSN 0013-5194. 57

Hecht, E., 2002, *Optics* (Addison Wesley), fourth edition. 15, 37

Hwang, W.-Y., 2003, Phys. Rev. Lett. **91**(5), 057901. 47

Jaduszliwer, B., 2008, Private conversation. 76

Jones, R. C., 1941, J. Opt. Soc. Am. **31**(7), 488, URL `http://www.opticsinfobase.org/abstract.cfm?URI=josa-31-7-488`. 18

Kranakis, E., 1986, *Primality and Cryptography* (Wiley). 5

Kurtsiefer, C., P. Zarda, S. Mayer, and H. Weinfurter, 2001, Journal of Modern Optics **48**, 2039. 35

Kwiat, P. G., E. Waks, A. G. White, I. Appelbaum, and P. H. Eberhard, 1999, Phys. Rev. A **60**(2), R773. 50

Lamas-Linares, A., and C. Kurtsiefer, 2007, Opt. Express **15**(15), 9388, URL `http://www.opticsexpress.org/abstract.cfm?URI=oe-15-15-9388`. 35

Lee, K. F., J. Chen, C. Liang, X. Li, P. L. Voss, and P. Kumar, 2006, Opt. Lett. **31**(12), 1905, URL `http://ol.osa.org/abstract.cfm?URI=ol-31-12-1905`. 53

van Lint, J. H., 1982, *Introduction to Coding Theory* (Springer). 39

Loudon, R., 1973, *The Quantum Theory of Light* (Oxford). 45

Lutkenhaus, N., and M. Jahma, 2002, New Journal of Physics **4**, 44, URL `http://stacks.iop.org/1367-2630/4/44`. 45, 46

Ma, L., H. Xu, and X. Tang, 2006 (SPIE), volume 6305, p. 630513, URL `http://link.aip.org/link/?PSI/6305/630513/1`. 67

MacWilliams, F. J., and N. J. A. Sloane, 1978, *The Theory of Error-Correcting Codes* (North-Holland). 42

Marcikic, I., A. Lamas-Linares, and C. Kurtsiefer, 2006, Applied Physics Letters **89**(10), 101122 (pages 3), URL `http://link.aip.org/link/?APL/89/101122/1`. 65

Matthews, J. C. F., A. Politi, A. Stefanov, and J. L. O'Brien, 2009, Nat Photon **3**(6), 346, URL `http://dx.doi.org/10.1038/nphoton.2009.93`. 8

McClelland, J. J., and J. L. Hanssen, 2006, Physical Review Letters **96**(14), 143005 (pages 4), URL `http://link.aps.org/abstract/PRL/v96/e143005`. 54

Migdall, A. L., D. Branning, and S. Castelletto, 2002, Phys. Rev. A **66**(5), 053805. 49

National Bureau of Standards, 1977, FIPS-Pub. 46, Data Encryption Standard. 12

Ohanian, H. C., 1986, American Journal of Physics **54**, 500. 26

Peev, M., C. Pacher, R. Alleaume, C. Barreiro, J. Bouda, W. Boxleitner, T. Debuisschert, E. Diamanti, M. Dianati, J. F. Dynes, S. Fasel, S. Fossier, *et al.*, 2009, New Journal of Physics **11**(7), 075001 (37pp), URL http://stacks.iop.org/1367-2630/11/075001. 75

Peng, C.-Z., J. Zhang, D. Yang, W.-B. Gao, H.-X. Ma, H. Yin, H.-P. Zeng, T. Yang, X.-B. Wang, and J.-W. Pan, 2007, Physical Review Letters **98**(1), 010505 (pages 4), URL http://link.aps.org/abstract/PRL/v98/e010505. 48

Pomerance, C., J. W. Smith, and R. Tuler, 1988, SIAM Journal on Computing **17**(2), 387, URL http://link.aip.org/link/?SMJ/17/387/1. 2

Rarity, J. G., P. R. Tapster, P. M. Gorman, and P. Knight, 2002, New Journal of Physics **4**, 82, URL http://stacks.iop.org/1367-2630/4/82. 74

Reidler, I., Y. Aviad, M. Rosenbluh, and I. Kanter, 2009, Physical Review Letters **103**(2), 024102 (pages 4), URL http://link.aps.org/abstract/PRL/v103/e024102. 11

Rényi, A., 1961, in *Proc. 4th Berkeley Sympos. Math. Statist. and Prob., Vol. I* (Univ. California Press, Berkeley, Calif.), pp. 547–561. 40

Rivest, R. L., A. Shamir, and L. Adleman, 1978, Commun. ACM **21**(2), 120, ISSN 0001-0782. 2, 5

Rogers, D. J., J. C. Bienfang, A. Mink, B. J. Hershman, A. Nakassis, X. Tang, L. Ma, D. H. Su, C. J. Williams, and C. W. Clark, 2006 (SPIE), volume 6304, p. 630417, URL http://link.aip.org/link/?PSI/6304/630417/1. 63

Rogers, D. J., J. C. Bienfang, A. Nakassis, H. Xu, and C. W. Clark, 2007, New Journal of Physics **9**(9), 319, URL http://stacks.iop.org/1367-2630/9/319. 57

Schmitt-Manderbach, T., H. Weier, M. Fürst, R. Ursin, F. Tiefenbacher, T. Scheidl, J. Perdigues, Z. Sodnik, C. Kurtsiefer, J. G. Rarity, A. Zeilinger, and H. Weinfurter, 2007, Physical Review Letters **98**(1), 010504 (pages 4), URL http://link.aps.org/abstract/PRL/v98/e010504. 48, 64

Schneier, B., 1996, *Applied Cryptography* (Wiley). 41

Shannon, C. E., and W. Weaver, 1949, *A Mathematical Theory of Communication* (University of Illinois Press, Champaign, IL, USA), ISBN 0252725484. 2, 9

Shor, P., 1994, in *Foundations of Computer Science, 1994 Proceedings., 35th Annual Symposium on*, pp. 124–134. 3, 8

Singh, S., 2000, *The Code Book* (Anchor). 1

Swiss Quantum, 2009, Homepage, URL `http://www.swissquantum.com/`. 70

Takesue, H., E. Diamanti, C. Langrock, M. M. Fejer, and Y. Yamamoto, 2006, Opt. Express **14**(20), 9522, URL `http://www.opticsexpress.org/abstract.cfm?URI=oe-14-20-9522`. 55

Takesue, H., S. W. Nam, Q. Zhang, R. H. Hadfield, T. Honjo, K. Tamaki, and Y. Yamamoto, 2007, Nat Photon **1**(6), 343, URL `http://dx.doi.org/10.1038/nphoton.2007.75`. 60

Townsend, P., A. Ougazzaden, and P. Tapster, 1993, Electronics Letters **29**(14), 1291, URL `http://link.aip.org/link/?ELL/29/1291/1`. 68

Uchida, A., K. Amano, M. Inoue, K. Hirano, S. Naito, H. Someya, I. Oowada, T. Kurashige, M. Shiki, S. Yoshimori, K. Yoshimura, and P. Davis, 2008, Nat Photon **2**(12), 728, URL `http://dx.doi.org/10.1038/nphoton.2008.227`. 11

VanDevender, A. P., and P. G. Kwiat, 2007, J. Opt. Soc. Am. B **24**(2), 295, URL `http://josab.osa.org/abstract.cfm?URI=josab-24-2-295`. 59, 60

Vernam, G., 1926, Journal of the American Institute of Electrical Engineers **XLV**, 109. 1

Villoresi, P., T. Jennewein, F. Tamburini, M. Aspelmeyer, C. Bonato, R. Ursin, C. Pernechele, V. Luceri, G. Bianco, A. Zeilinger, and C. Barbieri, 2008, New Journal of Physics **10**(3), 033038 (12pp), URL `http://stacks.iop.org/1367-2630/10/033038`. 74

Wayne, M., E. Jeffrey, G. Akselrod, and P. Kwiat, 2009, Journal of Modern Optics **56**, 516. 29

Wheeler, J. A., 1999, in *Feynman and computation: exploring the limits of computers* (Perseus Books, Cambridge, MA, USA), ISBN 0-7382-0057-3, pp. 309–336. 40

Wiesner, S., 1983, SIGACT News **15**(1), 78, ISSN 0163-5700. 31

Wootters, W. K., and W. H. Zurek, 1982, Nature **299**(5886), 802, URL `http://dx.doi.org/10.1038/299802a0`. 33

Yoshizawa, A., R. Kaji, and H. Tsuchida, 2003, Electronics Letters **39**(7), 621, URL `http://link.aip.org/link/?ELL/39/621/1`. 53

Author's Biography

DANIEL J. ROGERS

Dr. Daniel J. Rogers is a staff scientist in the Applied Information Sciences Department at the Johns Hopkins University Applied Physics Laboratory in Laurel, Maryland. Prior to joining APL, Dr. Rogers worked at the National Institute of Standards and Technology developing technologies for broadband quantum key distribution. He holds a Doctorate in Chemical Physics from the University of Maryland and a Bachelor of Science in Physics and Mathematics from Georgetown University.